智能化运维实践
从Ansible到Kubernetes

吴文豪 孙靖翀 | 著

电子工业出版社
Publishing House of Electronics Industry
北京·BEIJING

内 容 简 介

本书主要介绍自动化运维和智能化运维的常用技术，同时，通过搭建实验环境的方式，让读者能够基于主流的容器化技术 Docker 与 Kubernetes 搭建自己的运维实验环境，从而帮助读者更好地掌握本书涉及的技术要点。

- 对于期望快速掌握容器化相关技术的读者，本书能够帮助读者快速完成 Docker 与 Kubernetes 的入门，迅速掌握容器化技术中常用的技术点，提升读者在容器化技术方面的能力。
- 对于期望掌握自动化运维技术的读者，本书详细介绍了自动化运维利器 Ansible 的使用方法，以及如何使用 Ansible 完成自动化运维中的日常任务。
- 对于期望掌握智能化运维技术的读者，本书介绍了一些"开箱即用"并且效果不俗的 AIOps 工具包，帮助读者快速掌握 AIOps 的关键工具与技术。

未经许可，不得以任何方式复制或抄袭本书之部分或全部内容。
版权所有，侵权必究。

图书在版编目（CIP）数据

智能化运维实践：从 Ansible 到 Kubernetes / 吴文豪，孙靖翀著. —北京：电子工业出版社，2021.5
（高效实战精品）
ISBN 978-7-121-41125-0

Ⅰ. ①智… Ⅱ. ①吴… ②孙… Ⅲ. ①程序开发工具 ②Linux 操作系统－程序设计 Ⅳ. ①TP311.561 ②TP316.85

中国版本图书馆 CIP 数据核字（2021）第 081727 号

责任编辑：陈晓猛
印　　刷：固安县铭成印刷有限公司
装　　订：固安县铭成印刷有限公司
出版发行：电子工业出版社
　　　　　北京市海淀区万寿路 173 信箱　　　邮编：100036
开　　本：787×980　1/16　　印张：15　　字数：336 千字
版　　次：2021 年 5 月第 1 版
印　　次：2025 年 2 月第 10 次印刷
定　　价：99.00 元

凡所购买电子工业出版社图书有缺损问题，请向购买书店调换。若书店售缺，请与本社发行部联系，联系及邮购电话：(010) 88254888，88258888。
质量投诉请发邮件至 zlts@phei.com.cn，盗版侵权举报请发邮件至 dbqq@phei.com.cn。
本书咨询联系方式：010-51260888-819，faq@phei.com.cn。

序

欣闻好友文豪重新开始写作，说是要把对 AI 和运维的理解和实践沉淀下来，作为这几年的工作注脚。回想一下，文豪的上一本著作《自动化运维软件设计实战》已是 2015 年的图书，距今已 6 年了，按照计算机行业的发展规律，已经算是"老黄历"了。

近几年云计算行业高速发展，新的技术、新的实践，层出不穷，我也有幸置身其中。

应用分布式化

随着 2014 年 10 月 7 日 Pivotal 发布第一个 Spring Cloud 的版本 1.0.0.M1 以来，凭借产品的易用性、良好的生态，Spring Cloud 迅速成为微服体系中最具代表性的开发框架，广大开发者在享受微服务开发所带来的便利同时，以前维护一个 Tomcat 的事情，现在起码都是 10 个微服务起步，这也给应用的运维引入了更大的复杂性。

Kubernetes 成为应用运行的标准平台

同样是在 2014 年，Google 将内部 Borg 系统第一次以开源的方式发布于 GitHub 之上，并将 Microsoft、Red Hat、IBM、Docker 引入 Kubernetes 社区。某种程度上，Kubernetes 重新定义了操作系统，应用通过 Kubernetes 定义的抽象层，能够享受传统架构下难以实现的自动资源调度、自动修复、水平扩缩容等能力，并提升了应用发布的质量，这是当年传统运维难以想象的，但是如何用好相关的能力，对运维工程师来说也是一个新的挑战。

传统监控升级提高了可观察性

几年前，我们手中的监控武器除了 Zabbix，还有一个不太成熟的 ELK，而现在，我们拥有

Prometheus、ELK Stack、SkyWalking、Zipkin、Grafana 等一系列工具。而且，我们已经看到了 OpenTelemetry 尝试从规范层面完成 Metric、Log、Trace 的大一统，困扰传统运维多年、多种运维数据难以关联的问题，即将得到解决。

AI 从"阳春白雪"变得触手可及

通常情况下，常见的 AI 技术面向的领域是视觉识别、NLP 等，如何将 AI 技术应用到运维领域，还是一个非常值得探索的问题。

看到了文豪新书的初稿，感觉本书来的正是时候，很好地体现了这几年运维的基础架构的技术发展，同时具备很强的动手指导性，能够帮助读者在实践的过程中，对相关的技术加深理解，为更深入地钻研相关技术打下基础。

期待文豪的新作能帮助大家走入云原生下的智能运维新时代。

陈自欣，运维产品专家

曾在 AIOps 领域创业，

目前就职于某头部互联网公司，

从事面向金融行业的分布式、

云原生等产品的商业化产品经理工作。

专家评价

近20年来，IT运维方式经历了从人工运维向工具运维和IT服务管理的发展，再向以云管为代表的自动化运维及正在爆发的智能化运维方向演进。本书两位作者分别是红帽自动化运维领域的资深专家和AIOps智能运维领域的产品研发专家，具有深厚的技术功底和丰富的行业落地实践经验。本书全面系统地对自动化和智能化运维方法进行了讲解，特别是展现了具体的操作实现方式，相信对读者具有很大启发和帮助，帮助运维人走向noOps之终极目标。

——田浩，IBM全球服务部高级服务经理，
中国移动自动化智能化运维能力成熟度模型设计顾问

IT运维从集中化走向自动化、智能化，是运维工作发展的必然，同时也要求运维工作者掌握相适应的运维技术和工具。作者在书中演示了利用Docker和Kubernetes技术搭建Ansible试验环境的全过程，并深入浅出地介绍了自动化运维和智能运维技术和工具，使读者可以通过实验操作更好地学习、掌握运维理论和技术，是相关专业学生和运维工作者不可多得的指导图书。

——曾波，网思科技副总裁兼数智化服务部总经理

数据中心近几年的基础架构发生了巨大的改变，从传统的IOE逐渐演进到灵活多变的混合云环境，这给IT运维工作带来了全新的挑战。本书的两位作者基于丰富的实战经验，深入浅出地讲解了如何高效迅速地实现运维的自动化和智能化部署，是运维技术人员不可多得的参考指南。

——熊志坚，网思科技高级副总裁（前IBM中国华南区信息系统服务部总经理）

第 46 次《中国互联网络发展状况统计报告》显示中国网民规模超过 9.40 亿，其中 99.20% 网民使用手机上网，同时相关的信息系统安全漏洞有 11073 个，较 2019 年同期（5853 个）增长 89.20%，其中高危漏洞有 4280 个，较 2019 年同期（1876 个）增长 128.10%。信息系统面临的运维挑战巨大。AIOps 系统概念从最初的 IOTA（IT Operations Analytics）升级为 AIOps（Algorithmic IT Operations），再发展到如今热门的 Artificial Intelligence for IT Operations，智能化运维是 IT 运营的未来。

本书立足于实际的 IT 运营工程实践，采用主流的开源运维工具，如 Kubernetes、Docker、Ansible，搭建自动化运维的实验场景。然后对 AIOps 工具包进行分类详尽的操作说明和配图展示，深入浅出地展示 AIOps 工具包与 AI 平台结合应用的落地实践。相信用心的读者，可以从中深入了解智能化运维领域的知识、技术和工具，完善自身的知识技能。

——胡大裟，四川大学计算机科学博士，美国休斯顿大学访问学者

随着软件架构复杂性的不断提升，运维的理念和技术手段也在不停地演进与发展，从早期的"人肉"运维，到今天大行其道的自动化运维，再到日渐完善的 AIOps，本书作者有幸经历了整个发展历程，由此积累了大量的一线实战经验。

如果你计划在 Kubernetes 环境下基于 Ansible 来学习自动化运维的实战知识，同时又想比较深入地理解 AIOps 的原理和实践，那么本书会是你的不二选择。

——茹炳晟，腾讯技术工程事业群基础架构部首席研发效能架构师，T4 级专家，腾讯研究院特约研究员

本书从当下运维的痛点入手，通过结合流行的容器平台很好地展示了在容器云的环境下如何使用 Ansible、AI 技术及其他社区项目来实现 AIOps，从而给广大 IT 运维团队提供了一个很好的迈向 AIOps 的思路，非常值得阅读。

——张亚光，红帽软件中国区解决方案部门架构师经理

前言

技术的更新迭代速度总是非常快，回想起编写《自动化运维软件设计实战》一书的时候，容器化技术还没有被广泛地使用，智能化运维的概念也还没有在运维圈火热起来。经过近几年的技术变迁，微服务、云原生、智能化运维等非常多的新技术和新概念陆续出现，并且获得了广泛应用。

新技术的出现，提升了运维工程师的工作效率。比如，在容器化技术出现之前，应用最终部署环境与测试开发环境的一致性问题是让运维工程师在完成应用部署时非常头疼的问题之一。在容器化技术出现之后，应用最终部署环境与测试开发环境的一致性问题被容器化技术完美解决了，运维工程师再也不需要为其担心了，而且由于使用了容器化技术，也提升了应用部署的效率。但是，事物往往存在两面性，新技术的出现虽然解决了不少问题，但也带来了新的问题。例如，容器化部署被广泛使用之后，容器的数量呈爆炸性增长，容器间调用的复杂性相较于传统部署模式的复杂性也数倍增加。因此，运维工程师需要为手中的运维工具箱增加一些更强劲的自动化运维和智能化运维工具，来应对新的技术浪潮。

开源社区中有非常多的运维工具包，所实现的功能及达到的效果参差不齐，本书选择了一些"开箱即用"并且效果不俗的开源工具包分享给读者。

本书章节内容如下。

第 1 章：

回顾自动化运维技术，介绍自动化运维过程中面临的问题，并且对自动化运维的后续发展进行展望，帮助读者快速了解自动化运维领域需要解决的问题及未来的发展方向。

第 2 章：

容器化技术被广泛应用之后，Kubernetes 技术的出现将容器化技术的普及推向了一个新的高度。本章主要介绍如何快速搭建 Kubernetes 实验环境，帮助读者快速掌握 Kubernetes 和 Docker 相关技术，为读者能够快速体验本书介绍的运维工具包提供了一套简单易用的实验环境。

第 3~4 章：

通过介绍 Ansible 的使用，以及采用 Ansible 实现自动化运维的典型案例，帮助读者掌握如何使用 Ansible 这款开源的自动化运维利器来完成日常运维工作。

第 5~7 章：

对智能化运维的发展历程进行了简单的回顾，并提供了对读者比较有帮助的 AIOps 工具包，以及介绍如何使用 Kubernetes 技术来搭建一个能够让 AIOps 技术快速落地的 AI 平台。

致谢

本书参考了大量的网络资料，这些资料来自 GitHub、Stack Overflow、知乎等，在此向这些促进知识传播的网络平台致以诚挚的敬意。

特别感谢我就职的网思科技股份有限公司，公司良好的技术氛围、快速成长的业务，让我有机会带领团队研发公司的拳头产品 AlphaMind AI 能力开放平台，这为本书的写作提供了非常好的外部环境。

感谢我的父母和妻子，以及我的女儿，你们在本书的写作过程中给予了我最大的支持。

最后，感谢各位读者朋友。

本书中的相关链接的地址可通过扫描封底二维码获取。

吴文豪

目录

第 1 章　自动化运维的常见问题与发展趋势 1

  1.1　运维过程中的常见问题 1

    1.1.1　设备数量多 1

    1.1.2　系统异构性大 2

    1.1.3　云计算技术成熟后带来更大的困难 2

    1.1.4　信息安全要求带来的挑战 3

  1.2　自动化运维主流工具 3

    1.2.1　SaltStack 4

    1.2.2　Ansible 5

  1.3　自动化运维 5

  1.4　新的趋势——AIOps 7

  1.5　小结 8

第 2 章　使用 Kubernetes 快速搭建实验环境 9

  2.1　Docker 10

    2.1.1　使用 Docker 搭建实验环境的优点 10

    2.1.2　安装 Docker 11

    2.1.3　Docker 的基础使用方法 16

- 2.1.4 Docker 常用命令与配置 17
- 2.1.5 定制 Ansible 镜像 18
- 2.1.6 使用 docker-compose 编排实验环境 20
- 2.1.7 docker-compose 的常用配置项 23
- 2.2 镜像仓库 24
  - 2.2.1 Docker Registry 24
  - 2.2.2 Harbor 26
- 2.3 Kubernetes 28
  - 2.3.1 Kubernetes 简介 28
  - 2.3.2 Kubeasz 29
  - 2.3.3 K3S 30
  - 2.3.4 Kubernetes 快速入门 30
  - 2.3.5 使用 Kubernetes Deployment 搭建 Ansible 实验环境 41

第 3 章 集中化运维利器——Ansible 59

- 3.1 Ansible 基础知识 59
  - 3.1.1 主机纳管——inventory 59
  - 3.1.2 动态 inventory 64
- 3.2 在命令行中执行 Ansible 65
  - 3.2.1 指定目标主机 66
  - 3.2.2 常用命令示例 67
- 3.3 Ansible 常用模块 68
  - 3.3.1 文件管理模块 69
  - 3.3.2 命令执行模块 73
  - 3.3.3 网络相关模块 76
  - 3.3.4 代码管理模块 77
  - 3.3.5 包管理模块 79
  - 3.3.6 系统管理模块 80
  - 3.3.7 文档动态渲染与配置模块 84

3.4 自动化作业任务的实现——Ansible Playbook ... 86
3.4.1 Playbook 示例 ... 86
3.4.2 常用的 Playbook 结构 ... 87
3.4.3 变量的使用 ... 91
3.4.4 条件语句 ... 95
3.4.5 循环控制 ... 96
3.4.6 include 语法 ... 98
3.4.7 Ansible Playbook 的角色 roles ... 99
3.5 密钥管理方案——ansible-vault ... 101
3.6 使用 Ansible 的 API ... 102
3.7 Ansible 的优点与缺点 ... 104

第 4 章 自动化运维 ... 105
4.1 Ansible 在自动化运维中的应用 ... 105
4.1.1 ansible_fact 缓存 ... 105
4.1.2 ansible_fact 信息模板 ... 107
4.1.3 载入 fact ... 107
4.1.4 set_fact 的使用 ... 108
4.1.5 自定义 module ... 108
4.2 挂载点使用情况和邮件通知 ... 109
4.2.1 任务目标 ... 109
4.2.2 任务分析 ... 109
4.2.3 任务的实现 ... 110
4.3 操作系统安全基线检查 ... 128
4.3.1 任务目标 ... 128
4.3.2 任务分析 ... 128
4.3.3 任务的实现 ... 129
4.4 收集被管理节点信息 ... 132
4.4.1 任务目标 ... 132

4.4.2 任务分析 132
4.4.3 Jinja2 简介 132
4.4.4 服务器巡检任务 140
4.5 小结 142

第 5 章 AIOps 概述 143
5.1 AIOps 概述 143
5.2 AIOps 的落地路线 144
5.3 基于基础指标监控系统的 AIOps 145
5.4 基于日志分析系统的 AIOps 149
5.5 基于知识库的 AIOps 151
5.6 基于 AI 平台的 AIOps 152

第 6 章 AIOps 工具包 154
6.1 应用系统参数自动优化 154
6.2 智能日志分析 162
    6.2.1 日志模式发现 163
    6.2.2 日志模式统计分析 170
    6.2.3 实时异常检测 171
6.3 告警关联分析 172
6.4 语义检索 178
    6.4.1 Bert-As-Service 178
    6.4.2 Bert Fine-tuning 180
6.5 异常检测 182
    6.5.1 典型场景——监控指标异常检测 183
    6.5.2 异常检测工具包——PyOD 183
6.6 时序预测 186
    6.6.1 典型场景——动态告警阈值 186
    6.6.2 时序预测工具包——Prophet 186

第 7 章 加速 AIOps 落地——AI 平台 189

7.1 AI 平台与 AIOps 189
7.1.1 为运维系统插上 AI 的翅膀 189
7.1.2 Polyaxon 190
7.2 搭建 AI 平台的技术点 195
7.2.1 nvidia-docker 196
7.2.2 nvidia-device-plugin 199
7.2.3 KubeShare——显卡资源调度 201
7.2.4 AI 算法插件框架设计 202
7.2.5 KEDA——基于事件的弹性伸缩框架 204
7.2.6 Argo Workflow——云原生的工作流引擎 211
7.2.7 Traefik 222
7.3 小结 225

第 1 章
自动化运维的常见问题与发展趋势

1.1 运维过程中的常见问题

1.1.1 设备数量多

在虚拟化技术发展起来之前,企业普遍是把单个应用部署在一台或多台服务器上进行 IT 建设的。随着公司内部的 IT 系统逐渐增加,运维工程师需要运维的服务器数量也随之增加。公司业务刚起步的时候,运维工程师可能只需要运维少量的服务器和业务系统,随着公司业务的发展,运维工程师需要运维的服务器和业务系统的数量往往会成倍增加。日常运维中有非常多重复性的工作,例如,为操作系统打补丁,对中间件、数据库等应用进行升级,纯手工的运维操作使得运维的效率非常低,也很容易因为非标准化的运维操作导致运维事故。

有人可能会问,为什么不把多个业务系统部署在一台服务器上呢?这样一方面可以减少由于服务器数量的增加而给运维工作带来的压力,另一方面可以为企业节省成本,关于这个问题,我们可以从以下几个角度进行分析。

首先,为了发挥各家供应商的长处,同时避免被某个软件厂商垄断了企业的业务系统,一一般情况下企业的业务系统都会交给不同的软件厂商进行开发。既然是不同的软件厂商，那么当多个业务系统被部署到一起时，很容易出现这样一个场景：A 公司和 B 公司的系统部署在同一台服务器上，A 公司于本周五增加了一个新功能，周末的时候出现了操作系统的 I/O 负载很高的情况，导致该服务器上的业务系统都出现了反应迟钝的现象。于是 B 公司的开发人员就开始抱怨 A 公司的新功能导致操作系统负载高，影响了他们系统业务的正常运行。A 公司的开发人员也回应道，他们的功能绝对没有问题，是 B 公司的系统周末出账太多才导致系统缓慢的，于是一场扯皮大战就这样开始了。

其次，从保证业务系统的可用性来看，不同的业务系统之间需要在操作系统层进行隔离。否则一旦操作系统出现问题，部署在操作系统上的所有业务系统都会出现故障，这肯定是达不到企业在业务系统可用性方面的要求的。而且，多套业务系统部署在一台服务器上，也会为性能优化、故障排查等后续处理带来许多干扰。

1.1.2　系统异构性大

给运维工程师带来困扰的第二个问题就是业务系统的异构性。

由于业务系统是由不同软件厂商开发的，不同软件厂商内部的技术栈会有差异，所以很容易出现 A 软件厂商开发的业务系统需要运行在 Red Hat 上，Web 服务器需要使用 Tomcat，数据库需要使用 MySQL，而 B 软件厂商开发的业务系统需要用 Windows Server 来承载，Web 服务器用的是 IIS、数据库用的是 SQL Server 的情况。

这时运维工程师就傻眼了，这怎么运维？系统出故障之后还得考虑究竟是用 SSH 去维护还是用远程桌面去维护。公司规定每月要有一次常规性的服务器重启，究竟哪个 IP 地址的服务器要用 SSH 去重启，哪个 IP 地址的服务器需要用远程桌面去重启？

1.1.3　云计算技术成熟后带来更大的困难

可能有一些运维工程师通过自己多年积累下的脚本战胜了系统异构性所带来的挑战，本以为可以歇一歇了，没想到近年来，随着云计算技术的日渐成熟，新的挑战来了。

以前企业的设备数量虽然会增长，但是毕竟需要经过一个漫长的企业内部流程才能完成设备的采购和上线。随着虚拟化技术的成熟，企业的 IT 建设再也不需要像以前一样，上线一个新的业务系统就得经历一个漫长的采购流程了，也不需要再费心思找机房放服务器了。IT 管理人员只需要申请一台虚拟机或一个容器，然后在 CMDB 里面填一下这台虚拟机或这个容器的信息就可以了。

说明：

CMDB（Configuration Management Database，配置管理数据库）用于存储与管理企业 IT 架构中设备的各种配置信息，它与所有服务支持和服务交付流程紧密相连，支持这些流程的运转、发挥配置信息的价值，同时依赖于相关流程保证数据的准确性。

在这个云计算技术普及的年代，IT 建设的成本在不断降低，IT 建设的速度也在不断提升，需要运维的设备数量从原来的几百台增加到几千台甚至上万台，而且很有可能这些也仅仅是这家企业的一个部门的设备数而已，这给运维工程师带来了更大的挑战。

1.1.4　信息安全要求带来的挑战

随着近年来国家层面对信息系统安全性的重视，企业也越来越意识到信息系统安全的重要性，这就需要运维工程师配合企业不断变化的安全审计工作，高频次地更新操作系统、应用系统的安全补丁，以及对应用系统进行多个层面的安全加固操作，传统的手工运维操作已经越来越难以应对了。如何在安全整改的过程中降低原始手工操作带来的各种不可控风险，保证变更过程中的系统一致性，成为运维工作的重点。

1.2　自动化运维主流工具

当服务器数量越来越多、服务器和业务系统的异构性越来越大的时候，有没有一种更加高效的方式来管理这些服务器呢？从竖井式 IT 建设到层次性 IT 建设的过程中，对于运维工程师来说，最大的问题是设备数量的爆炸性增长，原有靠堆人力或者积累脚本的做法已经显得不可行了。这时我们就需要思考这样一个问题，能不能有一个既可以不增加太多的人力成本，又可以可靠地批量完成集中化运维任务的方法呢？

在设备的架构比较一致的情况下，实现集中化运维并不是什么难事。对于 Windows 系统的设备，假如我们需要对它们进行集中化的重启，就可以采用 PowerShell 的方式对设备进行集中化的操作。例如，当需要对服务器进行重启的时候，可以通过 PowerShell 调用 WMI 的 API 来完成这个操作，而像一些 msi 程序的发布和安装，则可以通过 AD 域控的方式去完成，总体来说还是挺方便的。

在 Linux 和 UNIX 操作系统下，我们可以通过 SSH+Expect 的方式或者双机互信后采用 SSH 的方式完成集中化的运维，难度也不大。

但是，由于不同业务系统对可用性和鲁棒的要求不一样，通常会出现既有 Windows 系统的服务器，又有 Linux 和 UNIX 系统的服务器的场景。而这种操作系统的多样性，也给集中化的运维带来了不少麻烦。无论是 AD 域控+WMI 还是 SSH 的方式，都只是解决了一类服务器运维的问题。试想一下，现在需要对运维的所有服务器进行常规的集中化重启操作，我们还得先去看一下究竟哪些服务器是 Windows 操作系统，哪些服务器是 Linux 操作系统，再去做相应的操作。这种费时费力的操作方式并不是我们想要的，我们更希望有一个统一的入口，通过这个入口提供的对外封装好的一些接口对异构的服务器做集中化运维，如图 1-1 所示。

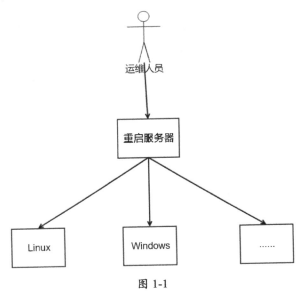

图 1-1

随着开源软件的不断发展，现在已经有许多不错的开源软件可供选择了，可以通过较低的成本实现集中化运维的目标。目前较为主流的开源集中化运维软件是 Ansible 和 SaltStack。

1.2.1 SaltStack

SaltStack 是一个用 Python 开发的集中化运维软件，它可以简化运维工程师对服务器的批量运维操作。SaltStack 内置了许多现成可用的模块，包括安装软件、配置参数、启停服务等功能。

SaltStack 支持的操作系统种类也十分丰富，它支持 Linux、UNIX、Solaris、Windows 等多种操作系统。

在软件的设计上，它支持 Master 主动推送配置和 Minion 定时拉取配置的方式，这一点与 Puppet 是十分类似的。同时，它还支持远程命令的并行执行，自带了许多日常执行模块，所以我们可以把 SaltStack 看作 Ansible 和 Puppet 的混合版本。SaltStack 还支持 SaltSSH 的方式，可以让我们无须使用 Agent 就能够对服务器轻易地进行批量操作。当我们希望 SaltStack 能够具备更好的扩展性，以及更好地使用 SaltStack 本身提供的模块时，我们可以在客户机上安装 Salt Minion 来进行服务器的集中化运维。

1.2.2 Ansible

Ansible 是一个用 Python 设计的无 Agent 模式集中化运维工具。以 SSH 的方式为主，支持多种远程连接的方式式对服务器进行集中化运维。也就是说，服务器上是不需要安装任何 Agent 端的；而针对 Windows 系统的服务器，可以使用 WRM（Windows Remote Management）实现无 Agent 模式管理。它与 SaltStack 非常类似，都是一种命令式的集中化运维工具。

虽然不需要安装任何 Agent 就可以使用，但是 Ansible 对被操作的服务器还是有一定要求的。例如，在调用一些模块的时候会提示客户机需要安装某个 Python 的扩展包。对于 Windows 系统的服务器来说，安装 PowerShell 3.0 或以上版本才可以让 Ansible 正常地运转。当然，Ansible 在 Windows 上需要将 PowerShell 作为任务执行的解释器。另外，Ansible 还可以通过对 VMWare、Docker、Kubernetes 等云产品的 API 支持实现对云主机的无 Agent 模式的支持。这样的做法也为我们提供了不少便利。

因此，本书后续部分主要基于 Ansible 讲解自动化运维的部分场景。

1.3 自动化运维

有了像 Puppet、Ansible 和 SaltStack 这样的工具，我们可以轻松地实现集中化运维了，一些集中化的部署、重启的操作都可以轻易完成，但是运维却还没有达到自动化的水平。例如，我们现在维护了 200 台设备，A 服务器上运行了一个业务系统，由于程序设计的原因，总是会不定期地出现应用服务器崩溃的问题，但是开发商由于技术问题一时解决不了，这时运维工程师就只能盯着自己的手机短信，一旦出现业务系统故障了，就得熟练地打开终端工具登录服务器，再熟练地输入 restart 命令。几天后，业务系统 B 无法上传文件了，运维工程师再次熟练地登录那台服务器，发现本来就不大的磁盘被系统写日志文件给写满了，于是运维工程师又熟练地输入 rm 命令删掉一些日志文件。一段时间后，又有用户报障……

运维工程师虽然有那么多集中化运维工具可以选择，但是却没有一款能帮助他们减轻负担。因为这些故障往往是非集中化的操作，解决的方法都是靠运维工程师在日常运维的过程中积累下来的特定经验，所以在这种情况下，无论使用哪一种集中化运维工具，和直接通过 SSH 或者远程桌面去操作其实没有多大的区别。问题出现一两次，我们可以登录出现故障的服务器进行处理，但是次数多了呢？难道我们就只有在每次故障出现之后重复做这些运维操作吗？

这时我们就需要把运维从集中化提升到自动化的水平了，而运维自动化，是一个对单点发生的故障运维知识沉淀的过程。

为什么说是单点发生？假如是多点发生并且重复的故障，我们通过集中化运维工具就可以很好地解决问题，但是面对单点发生的故障，集中化运维工具并不能产生很好的效果。

为什么说是知识的沉淀？假设出现服务器磁盘写满导致业务系统无法上传文件的故障，在排查后已经把问题定位清楚并且有解决方法了。对于运维工程师来说，这个过程就是一个经验沉淀的过程，处理故障后可以把这个处理过程中的经验知识积累下来，寻找一些方法使这个过程自动化。

对于这种由监控所驱动的自动化运维，一般由四个步骤组成，分别是了解、决策、执行，以及记录与反馈，如图 1-2 所示。

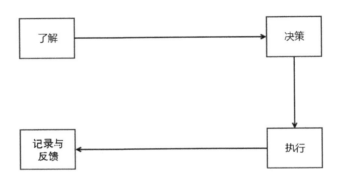

图 1-2

第一步是要对服务器、业务等我们所关注的对象进行监控。我们可以把监控看作一个了解信息的过程，只有当我们了解了实际情况之后，才能进行后续的操作。

第二步是决策，决策这个过程定义了在什么情况下我们应该执行什么操作的规则。例如，哪些服务器磁盘满了需要做删除日志文件的操作，哪些服务器磁盘满了需要做归档日志文件的操作，这些都是需要在决策过程中定义的。

第三步是执行，我们根据之前的规则做出相应的决策之后，就可以执行相应的运维操作了。在这个过程中我们需要关注的是如何屏蔽异构的操作系统带来的不便，也就是如何让异构操作系统的差异对执行的动作来说是透明的。

最后一步是记录和反馈，在完成具体的维护操作之后，我们需要有一套记录的机制，用于记录在什么时候做出了哪些运维操作，并且通过短信、邮件等方式给运维工程师发出通知，让运维工程师知道系统究竟执行了哪些运维操作，结果是什么。

对于了解、决策、记录和反馈这几个过程来说，目前开源软件中做得最好的就是 Zabbix 了，它具有丰富的告警策略、多种设备监控、强大的动作管理的功能，并且还具备了十分强大的扩展能力。而对于执行这个动作，可以让 Zabbix 与 Ansible 进行联动，当达到某个条件的时候根据规则触发对应的自动化运维脚本，从而达到自动化运维的效果。总的来说，通过监控系统和集中化运维工具的组合，能够将运维从批量化转为自动化，如图 1-3 所示。

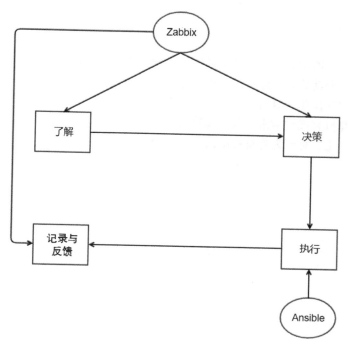

图 1-3

1.4 新的趋势——AIOps

AIOps（Artificial Intelligence for IT Operations，智能化运维）是将人工智能应用于运维领域，基于已有的运维数据（日志、监控信息、应用信息等），通过人工智能的方式进一步解决自动化运维无法解决的问题。

随着人工智能技术的扁平化，深度学习和机器学习技术的普及给原来自动化运维中无法实现的分析和预测功能提供了实现的可能。2016 年 Gartner 提出了 AIOps 的概念，自动化运维被纳入 AIOps 的专家系统子集。AIOps 不依赖于人为指定规则，主张由人工智能算法自动地从海量运维数据（包括事件本身及运维工程师的人工处理日志）中不断地学习，不断地提炼并总结规则，从而为运维工作增加了一个基于人工智能技术的大脑，指挥检测系统采集大脑决策所需的数据，在此基础上做出分析、预测，并指挥自动化运维系统去执行决策，从而实现运维系统的整体目标。

AIOps 是企业级 DevOps 在运维侧的高阶实现，是运维工作发展的必然趋势，是自动化运维的下一个发展阶段。

1.5 小结

没有一个非常好的定义能说明什么样的运维程度才能算得上自动化运维，有时我们认为能够批量地操作服务器就算自动化运维，能够定期出具巡检报告就算自动化运维，或者能够根据监控触发某些指定的操作就算自动化运维。笔者认为这些都没错，其实一切能让我们把人工的运维操作交给计算机完成的运维工具，都算得上自动化运维工具。不同的人对同一件事情会有相距甚远的看法，主要是因为各自接触的业务领域不同罢了。

第 2 章
使用 Kubernetes 快速搭建实验环境

为了更好地学习自动化运维及智能化运维工具,搭建一套实验环境是非常有必要的。本章介绍如何使用 Docker 和 Kubernetes 快速地搭建一套可重用的实验环境,并且会基于 Kubernetes 搭建:

- Ansible 实验环境;
- Prometheus Stack:用于采集基础监控指标,可作为智能化运维章节监控指标的数据来源。
- Loki:用于采集日志数据,可作为智能化运维章节日志的数据来源。

在容器化技术出现之前,一般采用在 VMWare 或者 KVM 中安装虚拟机的方式搭建实验环境,创建虚拟机后,通过虚拟机的链接克隆、快照等技术帮助我们快速地重置实验环境,从而达到进行可重试的实验的目的。

近些年,容器化技术日渐成熟,利用容器化技术搭建实验环境也是一个不错的选择。利用容器化技术,运维工程师能够更加快速地搭建实验环境,并且以更加少的资源启动更多的仿真环境。利用容器化技术搭建实验环境有以下优点:

(1)更高的实验环境搭建效率:在容器化环境下,可以利用 Dockerfile 对实验容器进行定制和管理,利用 Docker-Compose 或者 Kubernetes 技术进行实验环境的编排,快速完成基础实验环境的部署,在有限的资源下更加真实地仿真实验环境。镜像定制完成后,可以直接启动,需要重新进行实验的时候，可以对容器进行重启，新启动的容器实例会以定制镜像时的状态被启动，提高了实验环境搭建的效率。

（2）更多的实验环境：直接在虚拟机层面进行实验，每多一个新的实验环境，我们都需要开通新的虚拟机，但是利用容器化技术，我们只需要开通一个新的容器实例。相比虚拟机，容器显得更加轻量、高效，资源的占用率也更少。利用容器化技术，即使一台普通的笔记本电脑，也能够仿真比使用虚拟机更多的设备，让我们获得更好的实验体验。

虚拟机和容器的特性对比如表 2-1 所示。

表 2-1

特性	虚拟机	容器
启动时间	分钟级	秒级
硬盘使用	一般为 GB 级	一般为 MB 级
性能	弱于原生	接近原生
系统支持量	数十个单机	上千个单机
隔离级别	操作系统级	进程级
隔离策略	Hypervisor	CGroups

2.1 Docker

2.1.1 使用 Docker 搭建实验环境的优点

Docker 是一个让应用软件能够更加容易地被部署和运行的工具。通过 Docker，我们可以将软件和软件所需要的其他依赖打包在一起作为一个整体进行部署。利用容器化技术，我们不再需要担心软件所运行的环境和开发环境不一致的问题。对于实验环境来说，也是非常有帮助的，我们既可以更加便捷地分享自己的实验环境，也可以直接导入别人搭建好的实验环境，节省搭建实验环境的时间。

Docker 在某种程度上和虚拟机有点像，但它和虚拟机有非常多的不同，最主要的区别是 Docker 无须创建整个虚拟机，这个特性使得容器的大小远比虚拟机的大小要小，而且对资源的消耗也远比虚拟机要少。在这一点上，Docker 非常适合作为实验环境，我们能在一台普通的笔记本电脑上启动的容器数量远大于其能够创建的虚拟机数量。

Docker 的另外一个特性也使得 Docker 非常适用于搭建实验环境，每个 Docker 容器拥有与其他容器隔离的网络、磁盘、计算资源，看起来就像一台完备的虚拟机，并且在重启容器后，能够回滚到最初定制的状态。这个特性使得我们无须担心实验环境由于操作不当而损坏,可以大胆地进行各种实验操作。

2.1.2 安装 Docker

1. 以 root 用户安装 Docker

前提:读者已经准备好了实验用的 CentOS 7 操作系统。

编辑/etc/selinux/config 文件,关闭操作系统中的 SELinux。

```
SELINUX=disabled
```

同时在命令行中执行临时关闭 SELinux 的命令:

```
setenforce 0
```

操作完毕后,下载 Docker 二进制文件,并解压到/usr/local/bin 目录下:

```
tar -xvf ./docker-19.03.9.tgz
cp -Rf ./docker/* /usr/local/bin/
mkdir /etc/docker
```

解压完毕后,对 Docker 进行必要的配置,创建并编辑/etc/docker/daemon.json 文件,加入以下配置:

```
{
  "max-concurrent-downloads": 10,
  "log-driver": "json-file",
  "log-level": "warn",
  "log-opts": {
    "max-size": "10m",
    "max-file": "3"
  },
  "data-root": "/home/docker"
}
```

加入以上配置后,就可以通过 dockerd 命令启动 Docker 了:

```
dockerd
```

直接通过命令行启动 Docker 并不便于应用的管理。接下来，我们把 Docker 交给 systemd 进行管理：

```
cd /usr/lib/systemd/system
vi docker.service
```

提示：systemd 是一个系统管理守护进程、工具和库的集合，用于取代 system V 初始进程。systemd 的功能是集中管理和配置类 UNIX 系统。

在 docker.service 中加入以下配置：

```
[Unit]
Description=Docker Application Container Engine
After=network-online.target firewalld.service
Wants=network-online.target

[Service]
Type=notify
ExecStart=/usr/local/bin/dockerd
ExecReload=/bin/kill -s HUP $MAINPID
LimitNOFILE=infinity
LimitNPROC=infinity
LimitCORE=infinity
TimeoutStartSec=0
Delegate=yes
KillMode=process
Restart=on-failure
StartLimitBurst=3
StartLimitInterval=60s

[Install]
WantedBy=multi-user.target
```

检查 Docker 的状态：

```
systemctl daemon-reload
systemctl status docker
```

```
● docker.service - Docker Application Container Engine
   Loaded: loaded (/usr/lib/systemd/system/docker.service; disabled; vendor preset: disabled
)
   Active: inactive (dead)
     Docs: https://docs.docker.com

1月 31 19:45:22 localhost.localdomain systemd[1]: Starting Docker Application Containe.....
1月 31 19:45:22 localhost.localdomain dockerd[10649]: time="2021-01-31T19:45:22.6068428..."
1月 31 19:45:22 localhost.localdomain dockerd[10649]: time="2021-01-31T19:45:22.6352775..."
1月 31 19:45:22 localhost.localdomain dockerd[10649]: time="2021-01-31T19:45:22.6400520..."
1月 31 19:45:22 localhost.localdomain dockerd[10649]: time="2021-01-31T19:45:22.6405675..."
1月 31 19:45:22 localhost.localdomain dockerd[10649]: time="2021-01-31T19:45:22.6405916..."
1月 31 19:45:22 localhost.localdomain dockerd[10649]: time="2021-01-31T19:45:22.6406039..."
1月 31 19:45:23 localhost.localdomain systemd[1]: Started Docker Application Container...e.
1月 31 19:49:00 localhost.localdomain systemd[1]: Stopping Docker Application Containe...
1月 31 19:49:01 localhost.localdomain systemd[1]: Stopped Docker Application Container...e.
Hint: Some lines were ellipsized, use -l to show in full.
```

此时，Docker 服务已经被 systemd 管理了，但状态是未启动。接下来启动 Docker：

```
systemctl start docker
systemctl enable docker
```

```
[root@localhost ~]# systemctl enable docker
Created symlink from /etc/systemd/system/multi-user.target.wants/docker.service to /usr/lib/
systemd/system/docker.service.
```

此时，Docker 已经启动，并且加入了开机启动项，再次查看 Docker 的当前状态：

```
systemctl status docker
```

```
● docker.service - Docker Application Container Engine
   Loaded: loaded (/usr/lib/systemd/system/docker.service; enabled; vendor preset: disabled)
   Active: active (running) since 日 2021-01-31 19:45:23 CST; 3min 3s ago
     Docs: https://docs.docker.com
 Main PID: 10649 (dockerd)
    Tasks: 45
   Memory: 275.2M
   CGroup: /system.slice/docker.service
           ├─10649 /usr/local/bin/dockerd
           └─10660 containerd --config /var/run/docker/containerd/containerd.toml --log-l...

1月 31 19:45:22 localhost.localdomain systemd[1]: Starting Docker Application Containe.....
1月 31 19:45:22 localhost.localdomain dockerd[10649]: time="2021-01-31T19:45:22.6068428..."
1月 31 19:45:22 localhost.localdomain dockerd[10649]: time="2021-01-31T19:45:22.6352775..."
1月 31 19:45:22 localhost.localdomain dockerd[10649]: time="2021-01-31T19:45:22.6400520..."
1月 31 19:45:22 localhost.localdomain dockerd[10649]: time="2021-01-31T19:45:22.6405675..."
1月 31 19:45:22 localhost.localdomain dockerd[10649]: time="2021-01-31T19:45:22.6405916..."
1月 31 19:45:22 localhost.localdomain dockerd[10649]: time="2021-01-31T19:45:22.6406039..."
1月 31 19:45:23 localhost.localdomain systemd[1]: Started Docker Application Container...e.
Hint: Some lines were ellipsized, use -l to show in full.
```

此时 Docker 已经启动并且能够被正常使用了。

2. 非 root 账户使用 Docker 命令

安装 Docker 后,默认只允许 root 账户执行相关的操作,为了让其他用户能够使用 Docker 命令操作 Docker,我们需要新增一个 Docker 用户组,并将用户加入 Docker 用户组,使普通用户也能执行 Docker 相关命令。

```
groupadd docker
useradd alphamind
usermod -G docker alphamind
```

执行上述命令后,我们创建了一个名为 alphamind 的用户,并将 alphamind 用户加入 Docker 用户组。此时切换到 alphamind 用户,执行 Docker 的相关命令,发现非 root 用户也能成功操作 Docker 了。

```
su - alphamind
docker pull centos:7
```

```
CONTAINER ID        IMAGE               COMMAND             CREATED             STATUS
              PORTS               NAMES
[alphamind@VM-0-3-centos ~]$ docker pull centos:7
7: Pulling from library/centos
2d473b07cdd5: Pull complete
Digest: sha256:0f4ec88e21daf75124b8a9e5ca03c37a5e937e0e108a255d890492430789b60e
Status: Downloaded newer image for centos:7
docker.io/library/centos:7
```

尝试运行一个容器,发现其也可以正常运行:

```
docker run --rm -it --name alphamind-centos centos:7 bash
```

```
[alphamind@VM-0-3-centos ~]$ docker run --rm -it --name alphamind-centos centos:7 bash
[root@a4b020be6625 /]# ps -ef
UID         PID  PPID  C STIME TTY          TIME CMD
root          1     0  1 12:31 pts/0    00:00:00 bash
root         14     1  0 12:31 pts/0    00:00:00 ps -ef
[root@a4b020be6625 /]#
```

3. 以 Rootless 模式启动 Docker

能否使用非 root 模式启动 dockerd 进程呢?答案是肯定的,Docker 官方提供了以 Rootless 模式启动 dockerd 进程的方式。具体操作如下。

第一步,将 alphamind 用户加入 sudoer 列表,编辑/etc/sudoers.d/alphamind 文件,加入以下内容:

```
alphamind  ALL=(ALL) NOPASSWD:ALL
```

第二步,执行如下命令,为 rootless 的启动提供必要的环境:

```
cat <<EOF | sudo sh -x
cat <<EOT > /etc/sysctl.d/51-rootless.conf
user.max_user_namespaces = 28633
EOT
sysctl --system
EOF
```

执行完成后，切换到 alphamind 用户，就可以使用 Rootless 模式安装 Docker 了：

```
su - alphamind
curl -fsSL https://get.docker.com/rootless | sh
```

将以下代码加入 ~/.bashrc：

```
export XDG_RUNTIME_DIR=/home/alphamind/.docker/run
export PATH=/home/alphamind/bin:$PATH
export DOCKER_HOST=unix:///home/alphamind/.docker/run/docker.sock
```

第三步，启动 Docker，启动方式与 root 用户下启动 Docker 有所区别：

```
dockerd-rootless.sh --experimental
```

Docker 启动后，用 alphamind 用户尝试启动 CentOS 7 镜像，可以看到容器能够被正常启动：

```
docker run --rm -it --name alphamind centos:7 bash
```

```
[alphamind@VM-0-3-centos ~]$ docker run --rm -it --name alphamind centos:7 bash
[root@f87baa1405fb /]# ps -ef
UID        PID  PPID  C STIME TTY          TIME CMD
root         1     0  0 12:47 pts/0    00:00:00 bash
root        15     1  0 12:47 pts/0    00:00:00 ps -ef
```

接下来切换回 root 用户，运行 ps -ef|grep dockerd 命令，可以看到如下输出内容，证明 dockerd 进程已经是用 alphamind 用户运行的了：

```
[root@VM-0-3-centos ~]# ps -ef|grep dockerd
alphami+ 22401 21554  0 20:41 pts/0    00:00:00 rootlesskit --net=vpnkit --mtu=1500 --slirp4netns-sandbox=auto --slirp4netns-seccomp=auto --disable-host-loopback --port-driver=builtin --copy-up=/etc --copy-up=/run --propagation=rslave /home/alphamind/bin/dockerd-rootless.sh --experimental
alphami+ 22408 22401  0 20:41 pts/0    00:00:00 /proc/self/exe --net=vpnkit --mtu=1500 --slirp4netns-sandbox=auto --slirp4netns-seccomp=auto --disable-host-loopback --port-driver=builtin --copy-up=/etc --copy-up=/run --propagation=rslave /home/alphamind/bin/dockerd-rootless.sh --experimental
alphami+ 22437 22408  0 20:41 pts/0    00:00:03 dockerd --experimental
root     26559 25003  0 20:48 pts/2    00:00:00 grep --color=auto dockerd
```

然后执行 ps -ef|grep docker|grep -v dockerd 命令，可以看到启动的容器也是用 alphamind 用户运行的：

```
[root@VM-0-3-centos ~]# ps -ef|grep docker|grep -v dockerd
alphami+ 22450 22437  0 20:41 ?        00:00:00 containerd --config /home/alphamind/.docker/run/docker/containerd/containerd.toml --log-level info
alphami+ 31157 22927  0 20:56 pts/1    00:00:00 docker run --rm -it --name alphamind centos:7 bash
alphami+ 31193     1  0 20:56 ?        00:00:00 /home/alphamind/bin/containerd-shim-runc-v2 -namespace moby -id 78e3dd33b6b5d5e737a3fba85e34b31d8d061c9a637cf6918baa773b79cbc258 -address /home/alphamind/.docker/run/docker/containerd/containerd.sock
```

需要注意的是，Rootless 模式下运行的 Docker 是不能使用 1024 以下的端口号进行端口映射的。

2.1.3　Docker 的基础使用方法

接下来以 Ansible 为例，对 Docker 的基础使用方法进行介绍。

在启动 Docker 之前，需要下载或者定制一个 Docker 镜像，在 docker hub 上可以搜索我们需要的镜像。

以 CentOS 8 Docker 镜像为示例，启动容器后，在容器中安装 Ansible：

docker pull centos:8

```
8: Pulling from library/centos
7a0437f04f83: Pull complete
Digest: sha256:5528e8b1b1719d34604c87e11dcd1c0a20bedf46e83b5632cdeac91b8c04efc1
Status: Downloaded newer image for centos:8
docker.io/library/centos:8
```

查看已经下载的镜像：

docker images

```
REPOSITORY    TAG    IMAGE ID        CREATED       SIZE
centos        8      300e315adb2f    7 weeks ago   209MB
```

可以看到，本地已经下载了一个 CentOS 8 的 Docker 镜像。接下来启动一个容器实例：

docker run --rm -itd --name ansible centos:8

```
[root@localhost ~]# docker run --rm -itd --name ansible centos:8
bfd7710f07428f45897b73a4fe81178079d3539c1fd5fea57a062f25a386fcdd
```

容器启动后，可以通过 docker exec 命令以交互式的方式进入容器：

```
docker exec -it ansible bash
```

由于此容器是一个干净的 CentOS 8 容器,进入容器内部后,我们需要使用 yum 安装 Ansible。

```
yum install -y epel-release
yum install -y ansible
```

安装完成后,可以看到 Ansible 已经可用了:

```
[root@bfd7710f0742 /]# ansible
ansible              ansible-console      ansible-inventory    ansible-test
ansible-config       ansible-doc          ansible-playbook     ansible-vault
ansible-connection   ansible-galaxy       ansible-pull
```

按下 Ctrl+D 组合键,退出容器。退出容器后,输入以下命令:

```
docker stop ansible
```

通过 docker ps 命令查看正在运行的容器,发现容器已经被完全销毁:

```
[root@localhost ~]# docker ps
CONTAINER ID    IMAGE              COMMAND            CREATED           STATUS
      PORTS              NAMES
```

提示:由于启动容器的时候加入了 --rm 选项,所以容器停止运行后,容器会自动销毁,否则需要执行 docker rm ansible 命令才能将容器完全销毁。

2.1.4　Docker 常用命令与配置

掌握 Docker 的基本使用方法后,接下来快速了解一下 Docker 的常用配置。根据对实验环境的需求不同,可以对 Docker 进行自定义配置,从而构建一个更加适合自己的实验环境。dockerd 启动的时候,默认会读取 /etc/docker/daemon.json 配置文件,常用的配置项如表 2-2 所示。

表 2-2

配置项	说明
data-root	Docker 运行时的根路径,可以根据当前系统的磁盘大小进行合理的分配
insecure-registries	此配置项是一个数组,当自建容器镜像并且没有配置 HTTPS 的时候,需要配置此配置项,否则无法将镜像推送至镜像仓库
registry-mirrors	此配置项是一个数组,使用此配置项可以配置下载镜像时的地址,起到加速镜像下载的作用

除了上述的常用配置项,还有一些 Docker 的命令是在实验过程中经常用到的。

下载指定的镜像到本地：

```
docker pull centos:7
```

查看本地的镜像：

```
docker images
```

删除指定的镜像：

```
docker rmi <镜像 id>
```

删除所有名称为 none 的镜像：

```
docker images | grep none | awk '{print $3}' | xargs docker rmi
```

以交互式的模式进入容器：

```
docker exec -it <容器名称> sh
```

查看容器日志：

```
docker logs -f <容器 id>
```

2.1.5 定制 Ansible 镜像

将容器停止之后，再次启动容器，会发现容器中所有安装的 Ansible 应用不见了，这是 Docker 的特性。

当我们启动容器后，容器所处的环境位于镜像文件的最上层，此层会在退出或重启镜像后被重置，类似于给计算机装了还原卡一样，一旦计算机重启，所有的应用都会还原到最初的状态。有没有一个简单的方法让容器保持我们定制的状态，而不是每次重置容器后都要重新安装应用呢？答案是肯定的，我们可以通过编写 Dockerfile 定制属于自己的镜像，定制好的镜像会达到我们期望的最终状态。

以定制 Ansible 的镜像为例，新增以下内容到一份名称为 Dockerfile 的文件中：

```
FROM centos:7
WORKDIR /etc/yum.repos.d/
RUN yum install -y wget
RUN wget -O /etc/yum.repos.d/epel.repo http://mirrors.aliyun.com/repo/epel-7.repo
RUN yum clean all
RUN yum makecache
RUN yum install -y ansible openssh-server openssh-clients
RUN echo "123456"|passwd --stdin root
```

```
RUN ssh-keygen -t rsa -f /etc/ssh/ssh_host_rsa_key -N "" -q && \
ssh-keygen -t ecdsa -f /etc/ssh/ssh_host_ecdsa_key -N "" -q && \
ssh-keygen -t ed25519 -f /etc/ssh/ssh_host_ed25519_key -N "" -q
CMD /usr/sbin/sshd && tail -f /var/log/wtmp
```

执行 docker build 命令定制镜像：

```
docker build -t ansible .
```

命令执行完成后，一个定制好的 Ansible 镜像就被制作出来了，可以通过 docker images 命令查看镜像：

```
[root@localhost ~]# docker images
REPOSITORY    TAG       IMAGE ID        CREATED          SIZE
ansible       latest    66e76f4021bf    48 seconds ago   960MB
centos        8         300e315adb2f    7 weeks ago      209MB
centos        7         8652b9f0cb4c    2 months ago     204MB
```

我们试着启动 Ansible 镜像，看一下是否按照我们的期望把 Ansible 预先安装好了：

```
docker run --rm -it ansible bash
```

```
[root@localhost ~]# docker run --rm -it ansible bash
[root@89d3d53c0277 C]# ansible
usage: ansible [-h] [--version] [-v] [-b] [--become-method BECOME_METHOD]
               [--become-user BECOME_USER] [-K] [-i INVENTORY] [--list-hosts]
               [-l SUBSET] [-P POLL_INTERVAL] [-B SECONDS] [-o] [-t TREE] [-k]
               [--private-key PRIVATE_KEY_FILE] [-u REMOTE_USER]
               [-c CONNECTION] [-T TIMEOUT]
               [--ssh-common-args SSH_COMMON_ARGS]
               [--sftp-extra-args SFTP_EXTRA_ARGS]
               [--scp-extra-args SCP_EXTRA_ARGS]
               [--ssh-extra-args SSH_EXTRA_ARGS] [-C] [--syntax-check] [-D]
               [-e EXTRA_VARS] [--vault-id VAULT_IDS]
               [--ask-vault-pass | --vault-password-file VAULT_PASSWORD_FILES]
               [-f FORKS] [-M MODULE_PATH] [--playbook-dir BASEDIR]
               [-a MODULE_ARGS] [-m MODULE_NAME]
               pattern
```

提示：此次定制 Ansible 镜像使用的是 CentOS 7 的系统环境，而首次实验时使用的是 CentOS 8 的系统环境，在容器的世界中，切换不同的操作系统是非常便捷的，而在虚拟机的世界中，我们只能安装两套操作系统。

容器虽然定制好了，但上面的 Dockerfile 命令都代表什么意思呢？下面我们对 Dockerfile 的一些关键命令进行讲解，如表 2-3 所示。

表 2-3

命令	说明
FROM	指定使用的基础镜像，必须写在第一行
RUN	构建镜像时在镜像内执行的命令
ADD	将本地文件添加至容器中（假如文件是 tar 类型，则会自动解压），也能够将网络文件添加至容器中
COPY	将本地文件添加至容器中，不会自动解压，也不能访问网络资源
CMD	容器构建完成后，启动容器时执行的命令
ENV	用于设置容器的环境变量
WORKDIR	指定容器当前的工作目录

我们再回头看一下 Dockerfile 都做了什么工作：

（1）指定了基础镜像是 CentOS 7（FROM centos:7）。

（2）切换当前工作目录（WORKDIR/etc/yum.repos.d/）。

（3）安装 wget（RUN yum install -y wget）。

（4）下载国内 EPEL 源。

（5）清空 yum 元数据（RUN yum clean all）。

（6）重建 yum 元数据（RUN yum makecache）。

（7）安装 Ansible 及 SSH 服务，为后面的实验环境做准备（yum install -y ansible openssh-server openssh-clients）。

（8）配置 SSH 服务（RUN ssh-keygen -t rsa -f /etc/ssh/ssh_host_rsa_key -N "" -q && \ ssh-keygen -t ecdsa -f /etc/ssh/ssh_host_ecdsa_key -N "" -q && \ ssh-keygen -t ed25519 -f /etc/ssh/ssh_host_ed25519_key -N "" –q）。

（9）设置默认的启动命令，在没有配置的情况下默认启动 SSH 服务（CMD /usr/sbin/sshd && tail -f /var/log/wtmp）。

2.1.6　使用 docker-compose 编排实验环境

Ansible 的镜像已经定制好了，我们需要一台目标设备来让 Ansible 作为运维的目标主机，按照常规的方法，执行以下两条命令来完成这个目标：

```
docker run --rm --name os -itd ansible
docker run --rm --name ansible --link os:os -itd ansible
```

提示：通过 --link 选项，可以将容器与另外一个容器连接起来，使得两个容器的网络能够互通。

进入 Ansible 容器时，我们可以"ping"通名称为 os 的容器，证明此时的网络已经连通了：

```
[root@097060d98347 yum.repos.d]# ping os
PING os (172.17.0.2) 56(84) bytes of data.
64 bytes from os (172.17.0.2): icmp_seq=1 ttl=64 time=0.089 ms
64 bytes from os (172.17.0.2): icmp_seq=2 ttl=64 time=0.045 ms
64 bytes from os (172.17.0.2): icmp_seq=3 ttl=64 time=0.048 ms
```

虽然这样操作也可以完成搭建实验环境的目标，但是仍然存在一些不方便的地方——启动和暂停容器都需要执行多个命令。那么如何才能用一个简单便捷的方法使得多容器的实验环境搭建起来更加容易呢？此时我们可以使用 docker-compose。

docker-compose 是 Docker 官方的开源项目，负责实现对 Docker 容器集群的快速编排。也就是说，使用 docker-compose，我们能够快速地完成多个容器的启动与停止，让实验过程更加便捷。

```
docker-compose-`uname -s`-`uname -m` > /usr/local/bin/docker-compose
chmod +x /usr/local/bin/docker-compose
```

```
[root@localhost ~]# docker-compose
Define and run multi-container applications with Docker.

Usage:
  docker-compose [-f <arg>...] [--profile <name>...] [options] [--] [COMMAND] [ARGS...]
  docker-compose -h|--help

Options:
  -f, --file FILE             Specify an alternate compose file
                              (default: docker-compose.yml)
  -p, --project-name NAME     Specify an alternate project name
                              (default: directory name)
  --profile NAME              Specify a profile to enable
  -c, --context NAME          Specify a context name
  --verbose                   Show more output
```

提示：docker-compose 的安装包是一个二进制文件，下载后可直接使用。

安装完成后，将上面用命令启动的两个 Docker 容器转换为 docker-compose 的形式。

新建 docker-compose.yml 文件：

```yaml
version: "3"
services:
  os:
    image: ansible
    container_name: os
```

```
ansible:
  image: ansible
  container_name: ansible
  command:
    - /bin/bash
    - -c
    - "while true;do sleep 1;done"
```

执行 docker-compose up –d 命令：

```
[root@localhost ~]# docker-compose up -d
Building with native build. Learn about native build in Compose here: https://docs.docker.co
m/go/compose-native-build/
Creating network "root_default" with the default driver
Creating ansible ... done
Creating os      ... done
```

执行 docker-compose ps 命令：

```
Name           Command                          State    Ports
---------------------------------------------------------------
ansible   /bin/bash -c while true;do ...        Up
os        /bin/sh -c /usr/sbin/sshd ...         Up
```

可以看到，我们用一份配置文件同时让多个容器启动了，进入 Ansible 容器，尝试 "ping" os 容器：

```
[root@5245545501bc yum.repos.d]# ping os
PING os (172.19.0.2) 56(84) bytes of data.
64 bytes from os.root_default (172.19.0.2): icmp_seq=1 ttl=64 time=0.175 ms
64 bytes from os.root_default (172.19.0.2): icmp_seq=2 ttl=64 time=0.068 ms
64 bytes from os.root_default (172.19.0.2): icmp_seq=3 ttl=64 time=0.066 ms
```

可以发现，使用 docker-compose 启动的容器，网络已经自动连通了，通过 docker-compose 文件中的名称就可以互相访问。

执行 docker-compose down 命令，compose 文件中配置的容器和网络就同时被移除了：

```
[root@localhost ~]# docker-compose down
Stopping ansible ... done
Stopping os      ... done
Removing ansible ... done
Removing os      ... done
Removing network root_default
```

通过 docker-compose，不仅极大地提升了多容器实验环境的构建与销毁速度，同时还让实验环境的创建更加标准化。

2.1.7　docker-compose 的常用配置项

在 docker-compose 中，一个最小的配置文件如下：

```
version: "3"

services:
 ansible:
   image: ansible
```

配置文件中指定了我们需要启动的镜像。加入实验环境所需要的配置，参考如下。

- cap_add：用于为容器加入系统内核的能力

  ```
  cap_add:
    - ALL
  ```

- command：用于覆盖容器启动后的命令

  ```
  command:
    - /bin/bash
    - -c
    - "while true;do sleep 1;done"
  ```

- image：指定容器所使用的镜像名称

  ```
  image: ansible
  ```

- ports：声明容器暴露的端口信息

  ```
  - ports:
      - "22:22"
  ```

- volumes：声明数据卷的挂载路径

  ```
  volumes:
    - /data:/data
  ```

- restart：指定容器的重启策略

  ```
  restart: always
  ```

- tty：声明是否需要模拟一个伪终端

  ```
  tty: true
  ```

2.2 镜像仓库

在制作完 Docker 镜像之后，我们可以把实验环境存放在 Docker 的镜像仓库中进行统一管理，一方面可以达到快速恢复实验环境的目的，下次在一台新的设备上做实验的时候，就不需要再次编译镜像了，可以直接从镜像仓库中"pull"下来；另一方面可以对实验进行统一归档，下次需要启动实验环境的时候，从镜像仓库中"pull"下来即可。

2.2.1 Docker Registry

Docker Hub 是 Docker 官方用于管理公共镜像的地方，我们除了可以在上面找到想要的镜像，还可以把定制的镜像推送上去。但是，我们期望有一个私有的镜像仓库来管理实验的镜像。这时就可以采用 Registry 镜像来达到目的。

Registry 在 GitHub 上有两份代码：老代码库和新代码库。老代码库是采用 Python 编写的，存在 pull 和 push 的性能问题，在 0.9.1 版本之后就标志为 deprecated，不再继续开发。从 2.0 版本开始，就在新代码库中进行开发，新代码库是采用 Go 编写的，修改了镜像 id 的生成算法、Registry 上镜像的保存结构，大大优化了"pull"和"push"镜像的效率。

官方在 Docker Hub 上提供了 Registry 的镜像，我们可以直接使用该 Registry 镜像来构建一个容器，搭建实验环境使用的私有仓库服务。

（1）从镜像仓库中下载镜像。

```
docker pull registry
```

```
latest: Pulling from library/registry
0a6724ff3fcd: Pull complete
d550a247d74f: Pull complete
1a938458ca36: Pull complete
acd758c36fc9: Pull complete
9af6d68b484a: Pull complete
```

（2）创建持久化卷，并启动 Registry 容器。

```
docker volume create registry
docker run -itd -v registry:/var/lib/registry -p 5000:5000 --restart=always --name registry registry
```

提示：使用 docker volume 可以创建持久化卷，镜像仓库所管理的镜像是需要被持久化的，否则容器被销毁后，所有的数据都会随之销毁。

（3）启动完成后访问 http://192.168.199.161:5000/v2/进行验证，可以看到 Registry 已经正常启动了，如图 2-1 所示。

```
1    // 20210208114653
2    // http://192.168.199.161:5000/v2/
3
4  ▾ {
5
6    }
```

图 2-1

（4）由于我们没有配置 HTTPS，所以要把仓库地址加入 Docker 配置文件的 insecure-registries 选项中，修改/etc/docker/daemon.json，加入完毕后重启 Docker。

"insecure-registries": ["192.168.199.161:5000"],

（5）尝试推送 Ansible 镜像至 Registry。

```
docker tag ansible 192.168.199.161:5000/ansible
docker push 192.168.199.161:5000/ansible
```

```
The push refers to repository [192.168.199.161:5000/ansible]
b53d938df927: Pushed
b9077236be67: Pushed
ddfac8766e36: Pushed
907bcf8edf02: Pushed
dbd530d13d4f: Pushed
174f56854903: Pushed
latest: digest: sha256:d1e35e0da30bfdfa15a4b1dc41673096276e475782326a3fc8062103c2706ad7 siz
```

（6）访问 http://192.168.199.161:5000/v2/_catalog 查看推送情况，如图 2-2 所示。

```
1    // 20210208115119
2    // http://192.168.199.161:5000/v2/_catalog
3
4  ▾ {
5  ▾    "repositories": [
6          "ansible"
7       ]
8    }
```

图 2-2

经过上述步骤后，Ansible 镜像已经被推送到镜像仓库了。

2.2.2 Harbor

在部署完 Registry 之后，会发现这个镜像仓库和我们所期望的并不太一样。它没有任何便于使用的图形化界面，这使得相关操作并不方便。除了 Registry，Harbor 是一个更好的镜像仓库选择，虽然相比 Registry，它的组件明显多了不少，但提供了更加丰富的功能及更好的操作体验。

Harbor 的主要功能：

- 基于角色的访问控制；
- 基于镜像的复制策略；
- 图形化用户界面；
- 支持 AD/LDAP；
- 镜像删除和垃圾回收；
- 审计管理；
- RESTful API；
- 部署简单。

虽然 Harbor 提供了非常多的功能，但对于实验环境来说，易用性是非常重要的。

接下来，我们以 harbor-2.0.1 为例对 Harbor 进行部署。Harbor 下载完毕后，对压缩包进行解压，查看 harbor 根目录下的 harbor.yml 文件，对以下配置进行修改：

```
hostname: 192.168.199.161
http:
  port: 5000
#https:
  #port: 443
harbor_admin_password: aiops
```

- 将 hostname 修改为 Harbor 所在设备的 IP 地址；
- 在 http 选项中，将端口指定为 5000；
- 关闭配置中的 https 选项；
- 指定 harbor_admin_password 的初始值为 aiops，用来设定 Harbor 的初始密码。

配置完毕后执行 sh ./install.sh 命令：

```
[Step 5]: starting Harbor ...
Building with native build. Learn about native build in Compose here: https://
-native-build/
Creating network "harbor-201_harbor" with the default driver
Creating harbor-log      ... done
Creating harbor-db       ... done
Creating registry        ... done
Creating redis           ... done
Creating registryctl     ... done
Creating harbor-portal   ... done
Creating harbor-core     ... done
Creating nginx           ... done
Creating harbor-jobservice ... done
✔ ----Harbor has been installed and started successfully.----
```

当出现以上内容时，表示 Harbor 安装完毕，访问 http://192.168.199.161:5000 页面，如图 2-3 所示，表示此时 Harbor 已经启动完成。

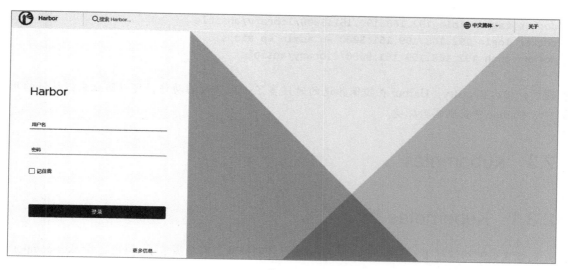

图 2-3

Harbor 的默认用户名为 admin，输入初始化配置的密码 aiops 后进入系统，可以看到 Harbor 的镜像管理界面，相比 Registry，Harbor 的功能更加强大，管理功能也更加易用，如图 2-4 所示。

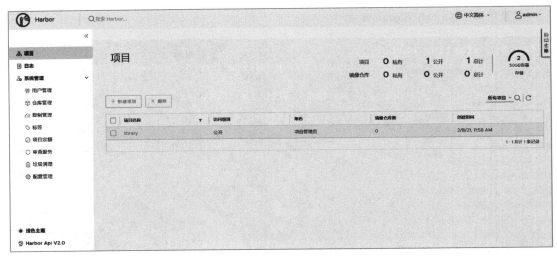

图 2-4

Harbor 可以使用后，重新设定镜像的 tag，并推送至 Harbor：

```
docker tag ansible 192.168.199.161:5000/library/ansible
docker login 192.168.199.161:5000 -u admin -p aiops
docker push 192.168.199.161:5000/library/ansible
```

提示：相比 Registry，Harbor 在镜像推送的时候多了一个登录的动作，可以防止未经认证的用户从 Harbor 中获取实验环境。

2.3 Kubernetes

2.3.1 Kubernetes 简介

在大多数的情况下，使用 docker-compose 已经能够满足实验需求了，但也有 docker-compose 所不能满足的场景：

- 期望搭建一个实验用的 PaaS 环境，将实验环境快速提供给多个实验者；
- 拥有多台实验设备，期望将这些设备资源充分地利用起来；
- 期望仿真环境能够提供负载均衡、网络隔离等特性，更好地对实验环境进行仿真。

在这种情况下，我们可以使用 Kubernetes 搭建更加高级的实验环境。

Kubernetes 是一个自动化部署、伸缩和操作应用程序（容器）的开源平台。使用 Kubernetes，可以快速、高效地满足以下需求：
- 快速精准地部署实验环境；
- 根据实验的需求弹性地伸缩实验环境；
- 提供网络隔离、资源限制、负载均衡等"开箱即用"的能力，更好地对实验环境进行仿真；
- 更加充分地利用资源；
- 提供多租户的能力将实验者的实验环境进行隔离。

2.3.2 Kubeasz

纯手工部署一套 Kubernetes 环境是非常麻烦的，也容易出错，使用 Kubeasz 对 Kubernetes 进行部署是一个不错的选择。Kubeasz 是一个基于 Ansible 的 Kubernetes 自动化部署项目，并且还考虑到了国内的网络环境，能够帮助我们快速地完成 Kubernetes 实验环境的部署。

- 下载 Kubeasz 的部署脚本
- 下载必需的安装文件

    ./ezdown -D

- 启动部署环境

    ./ezdown -S

- 单机快速部署

    docker exec -it kubeasz ezctl start-aio

- 检查节点是否正常工作

    kubectl get node

    ```
    NAME              STATUS   ROLES    AGE    VERSION
    192.168.199.161   Ready    master   129m   v1.20.2
    ```

- 检查系统组件是否正常工作

    kubectl get po -A

    ```
    [root@localhost ~]# kubectl get po -A
    NAMESPACE     NAME                                                  READY   STATUS    RESTARTS   AGE
    kube-system   coredns-5787695b7f-mntdl                              1/1     Running   0          107s
    kube-system   dashboard-metrics-scraper-79c5968bdc-vrvls            1/1     Running   0          94s
    kube-system   kube-flannel-ds-amd64-lc4r4                           1/1     Running   0          2m4s
    kube-system   kubernetes-dashboard-c4c6566d6-jm8pb                  1/1     Running   0          94s
    kube-system   metrics-server-8568cf894b-p557n                       1/1     Running   0          102s
    kube-system   node-local-dns-x4jw5                                  1/1     Running   0          106s
    ```

完成上述步骤后，一个 Kubernetes 的实验环境就已经搭建好了。

2.3.3 K3S

K3S 是经过 CNCF 认证，专为物联网及边缘计算而设计的 Kubernetes 发行版，体积小巧、资源占用更少，以及能够在 ARM 上运行都是 K3S 的亮点。K3S 官方提供了非常便捷的部署方式，对于希望能够快速搭建 Kubernetes 环境的读者来说，同样是一个非常不错的选择。相比 Kubernetes，K3S 的主要特点如下：

- 应用被打包为单个二进制文件，更加精简；
- 使用 SQLite3 作为默认的存储方案，但 etcd、MySQL、Postgres 依然可用；
- 最小化外部依赖。

K3S 的部署也非常简单，执行如下命令即可完成 K3S 单节点的快速部署：

```
curl -sfL http://rancher-mirror.cnrancher.com/k3s/k3s-install.sh |INSTALL_K3S_EXEC=
"--docker --kube-apiserver-arg service-node-port-range=1-65535" INSTALL_K3S_MIRROR=
cn sh -
```

安装完成后，执行 kubectl get node 命令，可以看到节点已经就绪：

```
[root@VM-0-3-centos ~]# kubectl get node
NAME             STATUS   ROLES                  AGE   VERSION
vm-0-3-centos    Ready    control-plane,master   98s   v1.20.5+k3s1
```

执行 kubectl get po -A 命令，可以看到 K3S 在安装的时候默认将 Traefik、Helm、local-path-provisioner 和 CoreDNS 都部署好了：

```
[root@VM-0-3-centos ~]# kubectl get po -A
NAMESPACE     NAME                                      READY   STATUS      RESTARTS   AGE
kube-system   coredns-854c77959c-42t98                  1/1     Running     0          114s
kube-system   metrics-server-86cbb8457f-9fq94           1/1     Running     0          114s
kube-system   helm-install-traefik-h8p65                0/1     Completed   0          114s
kube-system   local-path-provisioner-5ff76fc89d-fglpm   1/1     Running     0          114s
kube-system   svclb-traefik-bt78l                       2/2     Running     0          53s
kube-system   traefik-6f9cbd9bd4-qj7k7                  1/1     Running     0          53s
```

2.3.4 Kubernetes 快速入门

搭建好 Kubernetes 环境之后，我们需要学习如何操作 Kubernetes，以便更好地搭建实验环境。

1. Pod

1）Pod 的基本概念

Pod 是 Kubernetes 中可以创建的应用的最小单元，Pod 包含容器、存储、网络等的定义信息。有两种常见的 Pod 的使用方式：

（1）一个 Pod 中只运行一个容器。这是一种常见的用法，此时可以简单地认为 Pod 和容器是等价的。

（2）一个 Pod 中运行多个容器。当多个容器总是需要被部署在同一个节点，并且相关性比较高，需要通过 localhost 相互访问或者访问同一个本地卷时，一般会采用一个 Pod 中运行多个容器的方式。

Pod 中的容器可以共享两种资源：网络和存储。首先是网络的共享，每个 Pod 都会被分配唯一的 IP 地址。Pod 中的所有容器会共享 IP 地址和端口等网络资源。同一个 Pod 中的容器可以使用 localhost 互相通信。其次是存储的共享，可以为一个 Pod 指定共享的卷，使得 Pod 中的多个容器能够使用共享文件。共享卷的主要原因是 Pod 中的容器总是被同时调度到同一个节点，所以 Pod 中的各个容器看到的共享目录对应在主机上都是同一个目录。

2）Init 容器

InitContainers 是在 Kubernetes1.6 开始推出的一个新特性，Init 的配置方式和 Containers 的配置方式一致，只是各自的配置名称不一样，Init 容器配置在 initContainers 配置项中。它和普通的容器有以下不同的三点特性：

（1）Init 容器运行完成后就会终止。

（2）每个 Init 容器都必须在下一个 Init 容器启动成功之前运行完毕。

（3）如果 Pod 的 Init 容器启动失败，那么 Kubernetes 会不断地重启 Pod，直到 Init 容器启动成功为止。

Init 容器有哪些使用场景呢？

- 在 Init 容器中运行初始化工具，帮助应用做环境的初始化操作；
- 在设计软件的时候，可以将应用分为初始化服务和运行服务，对应用的职责进行划分；
- Init 容器使用 Linux Namespace，所以比普通容器能够获得更高的权限。

下面展示了 Init 容器的使用方法，在以下配置中，Pod 包含了一个 Nginx 容器，但这个 Nginx 容器的启动是有约束的，它需要 MySQL 服务启动完成之后才能够被启动，否则 Nginx 中所承载的应用会出现异常。此时我们就可以配置 Init 容器，让 Init 容器检查 MySQL 服务是否正常运行，只有正常运行才让 Nginx 服务启动，否则一直等待。

```yaml
apiVersion: v1
kind: Pod
metadata:
  name: app
  labels:
    app: app
spec:
  containers:
  - name: app
    image: nginx
    ports:
      - name: http
        containerPort: 80
        protocol: TCP
  initContainers:
    - name: init-mysql
      image: busybox:1.31
      command: ['sh', '-c', 'until nslookup mysql-svc; do echo waiting for mysql ……; sleep 2; done;']
```

3）Pause 容器

对于容器来说，Pod 其实是一个逻辑概念，真实运行的其实都是容器，每个启动的 Pod 必定伴随着一个启动的 Pause 容器。Pause 容器的功能只有一个，就是让自己永远处于 Pause 状态，它主要为 Pod 中的其他容器提供以下功能：

- 使得 Pod 中的不同应用能够看到其他应用的进程号；
- 使得多个容器能够访问同一个 IP 地址和端口范围；
- 使得多个容器能够通过 IPC 进行通信；
- 使得 Pod 中的多个容器共享一个主机名；
- 使得 Pod 中的多个容器能够访问在 Pod 级别定义的 Volumes。

总的来说，Pause 容器起到了桥梁的作用，解决了 Pod 中不同容器共享网络资源的问题。

4）Pod 的生命周期

和容器一样，Pod 也被认为是相对临时性的实体。Pod 在被创建后，会被赋予唯一的 ID，被调度到特定的节点，并在终止或删除之前一直在该节点上运行。如果 Pod 所在的节点宕机了，那么调度到该节点的 Pod 也会在一定时间后被删除。单纯的 Pod 是不具备故障自愈能力的，也就是说，如果 Pod 被调度到某个节点，由于节点故障而导致 Pod 被删除，那么 Pod 是不会自动重启或者调度到其他节点的。在 Kubernetes 中，负责调度 Pod 的组件被称作控制器，会在后续的章节中说明。

Pod 有自己的生命周期，了解 Pod 的生命周期能够帮助我们更好地找出 Pod 部署过程中失败的原因，了解当前集群中的 Pod 的运行是否健康。Pod 的运行状态如表 2-4 所示。

表 2-4

运行状态	描述
Pedding	Pod 已被 Kubernetes 系统受理，但有一个或者多个容器尚未创建亦未运行。此阶段包括等待 Pod 被调度和容器下载镜像两个环节
Running	Pod 已经绑定到了某个节点，Pod 中所有的容器都被创建。至少有一个容器仍在运行，或者正处于启动/重启状态
Succeeded	Pod 中的所有容器都成功运行，到达终止状态，并且不会再重启
Failed	Pod 中的所有容器都终止运行，并且至少有一个容器是因为失败而终止的。也就是说，容器以非 0 状态退出或者被系统终止

2. 集群管理

1）Node（节点）

当服务器加入 Kubernetes 之后，服务器就会成为 Kubernetes 中的一个 Node，可以通过 kubectl get node 命令查看当前节点的状态信息，包括节点当前的状态、角色、运行时长和版本信息。

```
[root@registry-svc ~]# kubectl get node
NAME            STATUS   ROLES    AGE   VERSION
192.168.30.50   Ready    master   32d   v1.18.3
```

对于 Node，有以下常用的运维命令：

（1）禁止 Pod 调度到指定节点。

```
kubectl cordon <节点名称>
```

（2）驱逐指定节点上的所有 Pod。执行该命令后，被驱逐的 Pod 会在其他节点上重新启动，一般在集群维护的时候会用到。

```
kubectl drain <节点名称>
```

2）NameSpace（命名空间）

Kubernetes 默认的设计是多租户模式，通过 NameSpace，可以让多个不同的用户或应用进行租户隔离，限定每个租户可用的资源。

（1）使用 NameSpace 隔离项目组：为不同的项目组创建 NameSpace，限定不同项目组的资源使用情况，项目组不需要使用资源的时候可以完成资源的快速回收。

（2）使用 NameSpace 隔离开发环境：为生产、测试、开发等环境划分不同的 NameSpace，方便对不同的环境进行管理。

有一点需要注意，并不是所有的资源对象都会有对应的 NameSpace，例如，Node 和 PersistentVolume 就不属于任何 NameSpace。NameSpace 既可以通过命令行创建，也可以使用配置文件的方式创建，使用配置文件创建的方式如下：

```yaml
apiVersion: v1
kind: Namespace
metadata:
  name: <NameSpace>
```

在创建命名空间的同时，可以限定命名空间的默认资源配额。例如，下面的配置在 NameSpace 中声明了 LimitRange，使得命名空间内的容器默认最多 CPU 使用数为 1 个，每个容器声明自己需要的 CPU 数量为 0.5 个。

```yaml
apiVersion: v1
kind: LimitRange
metadata:
  name: cpu-limit-range
spec:
  limits:
  - default:
      cpu: 1
    defaultRequest:
      cpu: 0.5
    type: Container
```

3）Taint 和 Toleration

Taint 和 Toleration 可以在 Node 和 Pod 上使用，主要的用途是优化 Pod 在集群间的调度方式。Taint 的作用是使节点能排斥一类特定的 Pod，Toleration 是配置在 Pod 上的，用于允许 Pod 被调度到与之匹配的 Taint 节点上。Taint 和 Toleration 相互配合，可以使 Pod 被分配到合适的运行节点上。

Taint 的操作命令如下，表示只有当拥有和这个 Taint 节点所匹配的 Toleration 时才允许被调度到此节点上：

```
kubectl taint nodes node1 key1=value1:NoSchedule
```

此时我们创建一个 Pod，为 Pod 加上 Tolerations 的配置，会发现这个 Pod 被调度到被标记为 Taint 的节点上：

```
apiVersion: v1
kind: Pod
metadata:
  name: nginx
  labels:
    env: test
spec:
  containers:
  - name: nginx
    image: nginx
    imagePullPolicy: IfNotPresent
  tolerations:
  - key: "key1"
    value: "value1"
    operator: "Exists"
    effect: "NoSchedule"
```

3. 控制器

前面在介绍 Pod 的时候讲到，Pod 自身是不具备故障自愈能力的，需要通过控制器才能获得故障自愈的能力。在实际使用中，我们很少直接管理 Pod，一般由控制器对 Pod 进行管理。常用的控制器有 Deployment、StatefulSet 和 DaemonSet。

1）Deployment

Deployment 控制器为 Pod 和 ReplicaSet 提供了一个声明式定义方法，用于替代以前的 ReplicationController 来方便地管理应用，它常用于创建无状态的应用。

以一个 Nginx 应用为例，编写如下 Deployment 文件。相比直接使用 Pod，使用 Deployment 配置能够告诉 Kubernetes 启动哪些应用，以及启动多少个副本，并且还能获得应用故障后自愈的特性。

```
apiVersion: apps/v1
```

```
kind: Deployment
metadata:
  name: nginx-deployment
  labels:
    app: nginx
spec:
  replicas: 3
  selector:
    matchLabels:
      app: nginx
  template:
    metadata:
      labels:
        app: nginx
    spec:
      containers:
      - name: nginx
        image: nginx:1.14.2
        ports:
        - containerPort: 80
```

当应用负载很高，需要扩容的时候，可以执行如下命令对应用进行扩容：

```
kubectl scale deployment nginx-deployment --replicas 10
```

除了手工指定副本数，还可以弹性扩容。例如，当 CPU 负载达到 80 以上的时候开始扩容，最小副本数为 10 个，最大副本数为 15 个：

```
kubectl autoscale deployment nginx-deployment --min=10 --max=15 --cpu-percent=80
```

2）StatefulSet

StatefulSet 和 Deployment 的最大区别是，StatefulSet 是有状态特性的，当我们要求 Pod 中的容器有序的时候，可以使用 StatefulSet。它具有如下特性：

- 唯一的网络标志；
- 持久化存储；
- 优雅地部署和伸缩；

- 优雅地删除和终止；
- 自动滚动升级。

下面以 Nginx 为例，配置一个有状态的 Nginx 服务，包含如下组件：

- 一个名称为 nginx 的 headless service；
- 一个名称为 web 的 StatefulSet，并且声明了 3 个运行 Nginx 容器的 Pod；
- 配置了 volumeClaimTemplates，给 Nginx 提供了稳定的存储。

代码如下：

```yaml
apiVersion: v1
kind: Service
metadata:
  name: nginx
  labels:
    app: nginx
spec:
  ports:
  - port: 80
    name: web
  clusterIP: None
  selector:
    app: nginx
---
apiVersion: apps/v1beta1
kind: StatefulSet
metadata:
  name: web
spec:
  serviceName: "nginx"
  replicas: 3
  template:
    metadata:
      labels:
        app: nginx
    spec:
      terminationGracePeriodSeconds: 10
      containers:
```

```yaml
    - name: nginx
      image: gcr.io/google_containers/nginx-slim:0.8
      ports:
      - containerPort: 80
        name: web
      volumeMounts:
      - name: www
        mountPath: /usr/share/nginx/html
volumeClaimTemplates:
- metadata:
    name: www
    annotations:
      volume.beta.kubernetes.io/storage-class: anything
  spec:
    accessModes: [ "ReadWriteOnce" ]
    resources:
      requests:
        storage: 1Gi
```

3）DaemonSet

DaemonSet 用于确保全部的 Node 都运行了一个给定的 Pod，当有新的节点接入集群时，会为它们新增一个指定的 Pod；当节点从集群中被移除时，这些 Pod 也会被回收。DaemonSet 有以下典型的用法：

- 集群存储 Daemon，如 Ceph、Glusterd；
- 日志采集 Agent，如 FileBeat、Nxlog 等；
- 基础监控 Agent，如 Zabbix、Node Exporter 等。

以 Node Exporter 为例，在使用 Prometheus 做基础监控的时候，经常会使用 Node Exporter 采集节点上的指标，此时可以使用 DaemonSet 提供的能力，让所有的节点都运行 Node Exporter，新加入的节点也自动运行 Node Expoter：

```yaml
apiVersion: apps/v1
kind: DaemonSet
metadata:
  name: node-exporter
spec:
  selector:
```

```yaml
      matchLabels:
        app: node-exporter
  template:
    metadata:
      labels:
        app: node-exporter
      name: node-exporter
      annotations:
        prometheus.io/scrape: 'true'
        prometheus.io/port: '9100'
        prometheus.io/path: '/metrics'
    spec:
      containers:
        - name: node-exporter
          image: registry-svc:25000/library/node-exporter
          imagePullPolicy: IfNotPresent
      hostNetwork: true
      hostPID: true
---
apiVersion: v1
kind: Service
metadata:
  name: node-exporter-svc
spec:
  selector:
    app: node-exporter

  ports:
  - protocol: TCP
    port: 9100
    targetPort: 9100
```

4. 存储管理

容器磁盘上的文件的生命周期是短暂的，默认情况下，当容器被重启后，容器中的文件就会被重置到镜像最初的状态，这对于有持久化需求的应用来说是不可接受的。因此，Kubernetes 提供了 Volume 来解决存储持久化的问题。Kubernetes 提供了非常多 Volume 的类型供我们使用。下面介绍常用的 Volume。

emptyDir：当 Pod 被分配给节点时，会创建 emptyDir 卷，并且只要 Pod 在节点上运行，该卷就会存在。当 Pod 被移除后，emptyDir 中的数据也会被删除。

```yaml
apiVersion: v1
kind: Pod
metadata:
  name: centos
spec:
  containers:
  - image: centos:7
    name: centos
    volumeMounts:
    - mountPath: /demo
      name: data
  volumes:
  - name: data
    emptyDir: {}
```

hostPath：用于将主机上的文件或目录挂载到集群中。

```yaml
apiVersion: v1
kind: Pod
metadata:
  name: centos
spec:
  containers:
  - image: centos:7
    name: centos
    volumeMounts:
    - mountPath: /data
      name: data
  volumes:
  - name: data
    hostPath:
      path: /data
      type: Directory
```

hostPath 类型的卷支持多种挂载的模式，通过 type 进行配置，常用配置如表 2-5 所示。

表 2-5

值	描述
DirectoryOnCreate	如果给定路径上什么都不存在,那么将根据需要创建空目录,权限设置为 0755,具有与 kubelet 相同的组和属主信息
Directory	在给定路径上必须存在的目录
FileOrCreate	如果给定路径上什么都不存在,那么将在给定的路径上根据需要创建空文件,权限设置为 0644,具有与 kubelet 相同的组和属主信息
File	在给定路径上必须存在的文件
Socket	在给定路径上必须存在的 UNIX 套接字
CharDevice	在给定路径上必须存在的字符设备
BlockDevice	在给定路径上必须存在的块设备

configMap:configMap 提供了向 Pod 注入配置数据的方法。在 configMap 对象中存储的数据可以被 configMap 类型的卷引用,然后被 Pod 中运行的容器化应用使用。引用 configMap 对象时,可以在 Volume 中通过它的名称来引用。

```
apiVersion: v1
kind: Pod
metadata:
  name: configmap-pod
spec:
  containers:
    - name: test
      image: busybox
      volumeMounts:
        - name: config-vol
          mountPath: /etc/config
  volumes:
    - name: config-vol
      configMap:
        name: log-config
        items:
          - key: log_level
            path: log_level
```

2.3.5 使用 Kubernetes Deployment 搭建 Ansible 实验环境

在 Kubernetes 环境中启动实验环境需要编写相关的配置文件,在开始配置之前,我们对

Kubernetes 的 Deployment 和 Service 配置方法进行简单的介绍。

1. Deployment 与 Service

Pod、Deployment、Service 的关系如图 2-5 所示。

在 Kubernetes 中，Pod 是最小的部署单元，Deployment 用于将多个 Pod 聚合起来，被 Deployment 聚合起来的 Pod 所暴露的端口能够使用 127.0.0.1 直接访问，Service 用于将 Deployment 的端口对外进行暴露，也能让不同 Deployment 之间的 Pod 通过 Service 的名称互相访问。总的来说，在 Kubernetes 的世界里，具备了很多 docker-compose 所不具备的高级特性，能够让我们的实验环境更加真实，具备更加完整的仿真体验，但也加大了实验环境管理的复杂性。

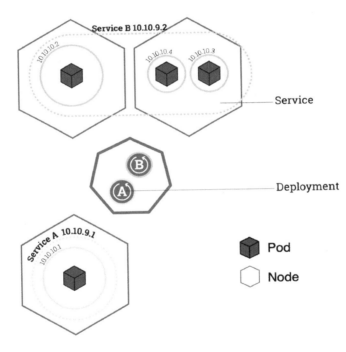

图 2-5

（1）常见的 Deployment 配置方式。

```
apiVersion: apps/v1
kind: Deployment                    # 指定资源的类型为 Deployment
metadata:                           # 配置资源的元数据
  name: ansible                     # 资源的名称
spec:
```

```yaml
  replicas: 2                       # 副本数量
  selector:                         # 配置标签选择器
    matchLabels:
      app: ansible
  template:
    metadata:
      labels:                       # 配置 Pod 的标签
        app: ansible
    spec:
      containers:
      - name: ansible               # 配置容器的名称
        image: ansible              # 配置容器所使用的镜像
        ports:
        - containerPort: 22         # 声明容器对外暴露的端口
```

（2）常见的 Service 配置方式。

```yaml
apiVersion: v1
kind: Service                       # 指定资源的类型为 Service
metadata:
  name: ansible-svc                 # Service 的名称
  labels:
    name: ansible-svc
spec:
  ports:
  - port: 22                        # 声明将 Service 的 22 端口映射到 Pod 的 22 端口
    targetPort: 22
    protocol: TCP
  selector:
    run: ansible                    # 声明此 Service 作用在 Ansible 上
```

2. 部署 Ansible 实验环境

编写以下内容并保存至 test.yml 文件中，然后执行 kubectl apply -f ./test.yml 命令启动实验环境：

```yaml
apiVersion: apps/v1
kind: Deployment
```

```yaml
metadata:
  name: ansible
spec:
  selector:
    matchLabels:
      app: ansible
  replicas: 1
  template:
    metadata:
      labels:
        app: ansible
    spec:
      containers:
        - name: ansible
          image: ansible
          imagePullPolicy: IfNotPresent
          command: ["/bin/sh"]
          args: ["-c", "while true; do sleep 10;done"]
---
apiVersion: apps/v1
kind: Deployment
metadata:
  name: centos
spec:
  selector:
    matchLabels:
      app: centos
  replicas: 1
  template:
    metadata:
      labels:
        app: centos
    spec:
      containers:
        - name: centos
          image: centos:7
          imagePullPolicy: IfNotPresent
```

```yaml
---
apiVersion: v1
kind: Service
metadata:
  name: centos-svc
spec:
  selector:
    app: centos
  ports:
  - protocol: TCP
    port: 22
    targetPort: 22
```

命令执行完毕后，检查容器是否正常运行：

```
kubectl get po
```

```
[root@localhost workspaces]# kubectl get po
NAME                          READY   STATUS    RESTARTS   AGE
ansible-69f4c66d44-zvpqn      1/1     Running   0          2s
centos-67fbbd69fc-wf8sn       1/1     Running   0          2s
```

当看到 Ansible 和 CentOS 的容器都处于 Running 状态时，实验环境的部署就完成了。通过 kubectl exec 命令进入容器内部：

```
kubectl exec -it ansible-69f4c66d44-zvpqn bash
```

进入容器内部后，执行 ping centos-svc 命令，可以看到在 Ansible 的容器内，通过 centos-svc 主机名称访问 CentOS 容器。此时，实验环境已经准备好了，接下来可以用 Ansible 操作名称为 CentOS 的目标容器：

```
[root@ansible-69f4c66d44-zvpqn yum.repos.d]# ping centos-svc
PING centos-svc.default.svc.cluster.local (10.68.127.189) 56(84) bytes of data.
64 bytes from centos-svc.default.svc.cluster.local (10.68.127.189): icmp_seq=1 ttl=64 time=0.080 ms
64 bytes from centos-svc.default.svc.cluster.local (10.68.127.189): icmp_seq=2 ttl=64 time=0.066 ms
```

添加双机互信：

```
ssh-keygen
```

```
Enter passphrase (empty for no passphrase):
Enter same passphrase again:
Your identification has been saved in /root/.ssh/id_rsa.
Your public key has been saved in /root/.ssh/id_rsa.pub.
The key fingerprint is:
SHA256:OyD/dRHCoSCvZk0TVQTQiHommb+XQaeGpnwYHZkXrUk root@ansible-69f4c66d44-dbc6h
The key's randomart image is:
+---[RSA 2048]----+
|    ..+*o++      |
|     .oEooo .    |
|    + +++. o .   |
|   = *+=.. . .   |
|    *==ooS .     |
|   .o=o+. .      |
|  . = o.oo .     |
|   + o o. o .    |
|    .  .  .      |
+----[SHA256]-----+
```

执行 ssh-copyid root@centos-svc 命令:

```
/usr/bin/ssh-copy-id: INFO: Source of key(s) to be installed: "/root/.ssh/id_rsa.pub"
The authenticity of host 'centos-svc (10.68.1.190)' can't be established.
ECDSA key fingerprint is SHA256:ju3iB4qFIoq2Gw8X2XBEyvULkw33S8cmoOG4UCd69AA.
ECDSA key fingerprint is MD5:87:c6:e2:f5:0f:cf:04:b4:26:f2:d5:82:81:29:45:35.
Are you sure you want to continue connecting (yes/no)? yes
/usr/bin/ssh-copy-id: INFO: attempting to log in with the new key(s), to filter out any that are already
 installed
/usr/bin/ssh-copy-id: INFO: 1 key(s) remain to be installed -- if you are prompted now it is to install
the new keys
root@centos-svc's password:

Number of key(s) added: 1

Now try logging into the machine, with:   "ssh 'root@centos-svc'"
and check to make sure that only the key(s) you wanted were added.
```

编辑/etc/ansible/hosts 文件，在末尾加入一行代码：

centos-svc

测试 Ansible 是否可用：

ansible all -m ping

```
[root@ansible-69f4c66d44-dbc6h yum.repos.d]# ansible all -m ping
centos-svc | SUCCESS => {
    "ansible_facts": {
        "discovered_interpreter_python": "/usr/bin/python"
    },
    "changed": false,
    "ping": "pong"
}
```

可用看到，此时已经能用 ansible 命令对 centos-svc 进行操作了。

3. 使用 Prometheus 搭建指标监控系统

无论是自动化运维还是智能化运维，都需要有基础的数据源，而时间序列数据是非常重要的数据来源之一，本节我们采用 Kubernetes 快速搭建一套 Prometheus 监控系统。

1）部署 Prometheus

整个部署文件比较长，主要还是用的 Deployment 和 Service 两个关键资源，但配置中多了 ConfigMap、ClusterRole、ServiceAccount 和 ClusterRoleBinding。其中，ConfigMap 定义了配置文件的内容，配置文件定义完成后，配置文件可以被挂载进入容器，这样也就使得配置项能够更加灵活地被定义。而配置 ClusterRole、ServiceAccount 和 ClusterRoleBinding 是因为 Prometheus 需要访问 Kubenetes 里面的资源，所以需要配置相应的账号并开通权限。

编写 config.yml 文件，定义 Prometheus 的配置文件：

```
apiVersion: v1
kind: ConfigMap
metadata:
  name: prometheus-config
  namespace: default
data:
  prometheus.yml: |-
    global:
      scrape_interval:     15s
      evaluation_interval: 15s
    scrape_configs:
    - job_name: 'kubernetes-pods'
      kubernetes_sd_configs:
      - role: pod
      relabel_configs:
      - source_labels: [__meta_kubernetes_pod_annotation_prometheus_io_scrape]
        action: keep
        regex: true
      - source_labels: [__meta_kubernetes_pod_annotation_prometheus_io_path]
        action: replace
        target_label: __metrics_path__
        regex: (.+)
      - source_labels: [__address__, __meta_kubernetes_pod_annotation_prometheus_io_port]
        action: replace
```

```
      regex: ([^:]+)(?::\d+)?;(\d+)
      replacement: $1:$2
      target_label: __address__
    - action: labelmap
      regex: __meta_kubernetes_pod_label_(.+)
    - source_labels: [__meta_kubernetes_namespace]
      action: replace
      target_label: kubernetes_namespace
    - source_labels: [__meta_kubernetes_pod_name]
      action: replace
      target_label: kubernetes_pod_name
---
```

编写 rbac.yml 文件，使 Prometheus 具备在 Kubernetes 集群内调用 Kubernetes 的能力：

```
apiVersion: rbac.authorization.k8s.io/v1beta1
kind: ClusterRole
metadata:
  name: prometheus
  namespace: default
rules:
- apiGroups: [""]
  resources:
  - nodes
  - nodes/proxy
  - services
  - endpoints
  - pods
  verbs: ["get", "list", "watch"]
- apiGroups:
  - extensions
  resources:
  - ingresses
  verbs: ["get", "list", "watch"]
- nonResourceURLs: ["/metrics"]
  verbs: ["get"]

---
```

```yaml
apiVersion: v1
kind: ServiceAccount
metadata:
  name: prometheus
  namespace: default
---
apiVersion: rbac.authorization.k8s.io/v1beta1
kind: ClusterRoleBinding
metadata:
  name: prometheus
roleRef:
  apiGroup: rbac.authorization.k8s.io
  kind: ClusterRole
  name: prometheus
subjects:
- kind: ServiceAccount
  name: prometheus
  namespace: default
```

编写 deployment.yml 文件,部署 Prometheus:

```yaml
---
apiVersion: apps/v1
kind: Deployment
metadata:
  name: prometheus
  namespace: default
spec:
  selector:
    matchLabels:
      app: prometheus
  replicas: 1
  template:
    metadata:
      labels:
        app: prometheus
    spec:
```

```yaml
      serviceAccountName: prometheus
      serviceAccount: prometheus
      volumes:
        - name: data
          hostPath:
            path: /home/data/nfs/prometheus
            type: DirectoryOrCreate
        - name: prometheus-config
          configMap:
            name: prometheus-config
      containers:
        - name: prometheus
          image: prom/prometheus
          imagePullPolicy: IfNotPresent
          volumeMounts:
            - name: data
              mountPath: /prometheus-data
              subPath: prometheus-data
            - name: prometheus-config
              mountPath: /etc/prometheus/
          ports:
            - containerPort: 9090
          command:
            - "/bin/prometheus"
          args:
            - "--config.file=/etc/prometheus/prometheus.yml"
            - "--web.external-url=http://127.0.0.1:9090/prometheus"
            - "--web.route-prefix=/prometheus"
---
apiVersion: v1
kind: Service
metadata:
  name: prometheus-svc
  namespace: default
spec:
  selector:
    app: prometheus
  type: NodePort
```

```
    ports:
    - protocol: TCP
      port: 9090
      targetPort: 9090
      nodePort: 9090
```

2）部署 node-exporter

node-exporter 能够采集系统的性能指标，如 CPU、内存、硬盘使用率等情况，由于部署 Prometheus 的时候配置了自动发现，所以部署 node-exporter 的时候，只需要启动应用即可，Prometheus 会自动发现并抓取指标。

```
apiVersion: apps/v1
kind: DaemonSet
metadata:
  name: node-exporter
spec:
  selector:
    matchLabels:
      app: node-exporter
  template:
    metadata:
      labels:
        app: node-exporter
      name: node-exporter
      annotations:
        prometheus.io/scrape: 'true'
        prometheus.io/port: '9100'
        prometheus.io/path: '/metrics'
    spec:
      containers:
        - name: node-exporter
          image: prom/node-exporter
          imagePullPolicy: IfNotPresent
      hostNetwork: true
      hostPID: true
---
apiVersion: v1
kind: Service
```

```yaml
metadata:
  name: node-exporter-svc
spec:
  selector:
    app: node-exporter

  ports:
  - protocol: TCP
    port: 9100
    targetPort: 9100
```

node-exporter 可以采集主机层面的指标，而容器的资源使用率可以通过 cadvisor 进行采集。和 node-exporter 一样，由于已经配置了自动发现，所以只要将 cadvisor 启动即可。

```yaml
apiVersion: apps/v1
kind: DaemonSet
metadata:
  name: cadvisor-exporter
spec:
  selector:
    matchLabels:
      app: cadvisor-exporter
  template:
    metadata:
      labels:
        app: cadvisor-exporter
      name: cadvisor-exporter
      annotations:
        prometheus.io/scrape: 'true'
        prometheus.io/port: '8080'
        prometheus.io/path: '/metrics'
    spec:
      containers:
        - name: cadvisor
          image: google/cadvisor
          imagePullPolicy: IfNotPresent
          volumeMounts:
            - name: rootfs
```

```yaml
          mountPath: /rootfs
        - name: var-run
          mountPath: /var/run
        - name: sys
          mountPath: /sys
        - name: docker
          mountPath: /var/lib/docker
        - name: disk
          mountPath: /dev/disk
      volumes:
        - name: rootfs
          hostPath:
            path: /
        - name: var-run
          hostPath:
            path: /var/run
        - name: sys
          hostPath:
            path: /sys
        - name: docker
          hostPath:
            path: /var/lib/docker
        - name: disk
          hostPath:
            path: /dev/disk
---
apiVersion: v1
kind: Service
metadata:
  name: cadvisor-exporter-svc
spec:
  selector:
    app: cadvisor-exporter

  ports:
  - protocol: TCP
    port: 8080
    targetPort: 8080
```

启动 Prometheus 后，访问 http://192.168.199.161:9090，单击 Status 菜单的 Targets 选项，可以看到监控客户端已经正常运行，如图 2-6 所示。

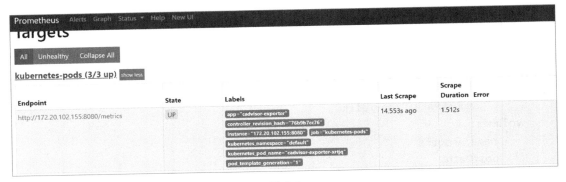

图 2-6

单击 Graph 菜单，可以查看各个指标项的情况，如图 2-7 所示。在以后的实验中，可以使用 Prometheus 获取时序指标作为实验用的数据源了。

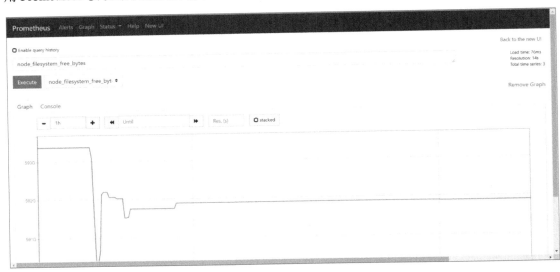

图 2-7

4. 使用 Loki 搭建日志采集系统

提到日志采集，读者肯定会想到 ELK（Elasticsearch、Logstash、Kibana 的首字母组合），ELK 是一套非常成熟可靠的日志采集和分析方案，但有些时候，我们并不需要对日志进行全文索引，并且我们期望以非常低的资源消耗完成对日志的采集、存储，以及简单的检索，这时 Loki 是一个不错的选择。

Loki 是 Prometheus 团队开源的一套支持水平扩展、高可用、多租户的日志聚合系统。它具有以下特点：

- 不对日志进行全文索引，存储的是经过压缩后的非结构化日志，只对索引元数据进行索引，具有非常明显的成本优势；
- Grafana 原生支持，提供了一套"开箱即用"的可视化 UI；
- 非常适合存储 Kubernetes Pod 的日志。

接下来搭建 Loki，首先安装 Promtail，Promtail 是专门为 Loki 设计的 Agent，负责收集日志并将其发送给 Loki。Promtail 的配置项较多，我们使用 Helm 安装 Promtail：

```
helm install promtail promtail --set "loki.serviceName=loki-svc"
```

Promtail 安装完成后，安装 Loki：

```
apiVersion: apps/v1
kind: Deployment
metadata:
  name: loki
spec:
  selector:
    matchLabels:
      app: loki
  replicas: 1
  template:
    metadata:
      labels:
        app: loki
    spec:
      containers:
      - name: loki
        image: grafana/loki
        ports:
        - containerPort: 3100
---
apiVersion: v1
kind: Service
```

```yaml
metadata:
  name: loki-svc
spec:
  selector:
    app: loki
  ports:
  - protocol: TCP
    port: 3100
    targetPort: 3100
```

最后部署 Grafana，查看 Loki 中存放的日志：

```yaml
apiVersion: v1
kind: Service
metadata:
  name: grafana-svc
spec:
  selector:
    app: grafana
  type: NodePort
  ports:
  - protocol: TCP
    nodePort: 3000
    port: 3000
    targetPort: 3000
---
apiVersion: apps/v1
kind: Deployment
metadata:
  name: grafana
spec:
  selector:
    matchLabels:
      app: grafana
  replicas: 1
  template:
    metadata:
```

```yaml
      labels:
        app: grafana
    spec:
      volumes:
      - name: data
        hostPath:
          path: /home/data/grafana
          type: DirectoryOrCreate
      containers:
      - name: grafana
        image: grafana/grafana
        volumeMounts:
        - name: data
          mountPath: /var/lib/grafana
          subPath: data
        ports:
        - containerPort: 3000
```

Grafana 部署完成后，需要登录 Grafana 配置 Loki 的数据源，如图 2-8 所示。

图 2-8

数据源配置完成后，即可进入 Explore 界面查看采集的日志，如图 2-9 所示。

图 2-9

第 3 章
集中化运维利器
——Ansible

3.1 Ansible 基础知识

3.1.1 主机纳管——inventory

Ansible 中的被管理节点需要以清单的方式保存在一个文件中，这个文件被称为 inventory 文件。默认情况下，Ansible 会自动搜索/etc/ansible/hosts 中的被管理节点清单。一个简单的被管理节点清单示例如下：

```
green.example.com
blue.example.com
192.168.100.1
192.168.100.10

[webservers]
alpha.example.org
beta.example.org
```

```
192.168.1.100
192.168.1.110

[dbservers]
db01.intranet.mydomain.net
db02.intranet.mydomain.net
10.25.1.56
10.25.1.57

db-[99:101]-node.example.com
```

第 1~4 行声明了 4 台未分组管理的被管理节点，通过示例可以看出被管理节点配置成 IP 地址或域名均可。

从第 6 行开始使用了两个特殊章节标志[webservers]和[dbservers]，这个章节标志在 Ansible 中称为分组（Groups），类似于 Windows 中的 ini 配置文件方式——将信息用[Group Name]的格式进行分组，分组后的被管理节点可以通过指定组名的方式来达到访问特定的被管理节点的目的。

注意：
如果被管理节点包含控制节点自身，则需要在 inventory 文件相应的主机名或地址中添加行为参数 ansible_connection=local，告知 Ansible 不需要通过 SSH 连接而是直接访问本地主机，例如：

```
127.0.0.1 ansible_connection=local
## 或者 ##
localhost ansible_connection=local
```

建议：
在构建一个 Ansible 项目时，可以将 inventory 文件放置在 Ansible 项目目录或 Ansible 项目的 inventory 子目录下。

1. 行为参数

Ansible inventory 文件中可使用的行为参数如表 3-1 所示。

表 3-1

参数名称	默认值	描述
ansible_host	无	SSH 访问的主机名或 IP 地址

续表

参数名称	默认值	描述
ansible_port	22	SSH 访问的目标端口
ansible_user	root	SSH 登录使用的用户名
ansible_password	无	SSH 登录所使用的密码
ansible_connection	smart	Ansible 使用何种连接模式连接到被管理节点
Ansible_private_key_file	无	SSH 使用的私钥
ansible_shell_type	sh	命令所使用的 shell
ansible_python_interpreter	/usr/bin/python	被管理节点上的 Python 解释器路径
ansible_*_interpreter	无	非 Python 实现的自定义模块使用的语音解释器路径

以上行为参数大体上都可以通过描述文件理解它的基本含义，但有几个行为参数需要额外说明。

1）ansible_shell_type

前文中已经提过 Ansible 支持多种传输机制，这里的传输机制具体是指 Ansible 连接到被管理节点的机制。默认值 smart 表示智能传输模式。智能传输模式会检测本地安装的 SSH 客户端是否支持一个名为 ControlPersist 的特性。如果 SSH 本地客户端支持该特性，那么 Ansible 将使用本地 SSH 客户端；如果本地 SSH 客户端不支持该特性，那么 Ansible 将使用一个名为 Paramiko 的 Python SSH 客户端（对于老版本的 OpenSSH 客户端的兼容性更好）。

2）ansible_shell_type

Ansible 建立远程连接后，默认在被管理节点上使用路径为/bin/sh 的 Bourne shell，并且会生成适用于 Bourne shell 的命令行环境参数。但是 ansible_shell_type 还可使用 sh、fish 和 powershell 作为该参数的合法值。

3）ansble_python_interpreter

该参数调用被管理节点上的/usr/bin/python 作为默认的 Python 解释器，但是很多版本的 Linux 操作系统都已经使用 Python 3 作为系统默认的 Python 解释器，这时就需要修改这个参数为 Python 3 解释器所在的路径，便于 Ansible 调用被管理节点上的 Python 3 解释器作为 Ansible 模块在远端执行的任务解释器。

4）ansible_*_interpreter

如果实现了非 Python 的自定义模块，则可以使用这个参数来指定解释器的路径（例如，/usr/bin/perl）。

2. 组

在实际工作中，我们希望将任务内容相同的被管理节点放在一起管理并执行任务，这就是 inventory 文件的分组功能。Ansible 默认实现了一个分组 all（或*），这个分组包括 inventory 文件中的所有被管理节点。前面已经提到，inventory 文件支持自定义分组，Ansible inventory 文件是.ini 格式的，在.ini 格式中使用[ ]将同类的设置值归集到一起成为一个分组。前面已经展示了 inventory 文件分组功能的一种实现方式，下面使用另外一种分组实现方式——先列出主机，再将它们加入特定的组中。示例如下：

```
green.example.com
blue.example.com
192.168.100.1
192.168.100.10
vagrant1 ansible_host=127.0.0.1 ansible_port=2200
vagrant2 ansible_host=127.0.0.1 ansible_port=2201
vagrant3 ansible_host=127.0.0.1 ansible_port=2202

[vagrant]
vagrant1
vagrant2
vagrant3
```

3. 别名

前面的 inventory 文件定义中使用了别名，我们复习一下：

```
[vagrant]
vagrant1 ansible_host=127.0.0.1 ansible_port=2200
vagrant2 ansible_host=127.0.0.1 ansible_port=2201
vagrant3 ansible_host=127.0.0.1 ansible_port=2202
```

这里的 vagrant1、vagrant2 和 vagrant3 就是别名，在一些特殊场景必须使用别名。例如，这里 inventory 文件所演示的情况，三台 vagrant 虚拟机在同时使用同一个 IP 地址不同的端口的情况下，需要使用别名对主机加以区分。如果这里不使用别名，则采用如下写法：

```
[vagrant]
127.0.0.1:2200
127.0.0.1:2201
127.0.0.1:2202
```

在执行 Ansible 命令或者 playbook 的时候，Ansible 只会操作 127.0.0.1:2202 这个被管理节点，并且会忽略 IP 地址为 127.0.0.1:2200 和 127.0.0.1:2201 的两台主机。

注意：
产生这个问题的原因是在被管理节点没有使用别名的情况下，Ansible 只能以现有的 IP 地址或域名作为标识符，相同的 IP 地址，由于加载的先后顺序，只能留下 127.0.0.1:2202 这一个有效的被管理节点信息，前面设置的两个被管理节点的信息被覆盖了。

4. 被管理节点的序列化描述

前面的 inventory 文件中有一个特别的主机描述：db-[99:101]-node.example.com。在 inventory 文件中，这样的配置描述了域名从 db-99-node.example.com 到 db-101-node.example.com 的 3 台被管理节点主机的信息。这种配置描述方式非常适合主机的 IP 地址或域名有规律的场景。例如，我们需要将 192.168.0.100 到 192.168.0.200 这个网段中所有的主机进行纳管，就可以在 inventory 文件中配置描述信息为 192.168.0.[100:200]。除这两种序列化描述方式外，常见的序列化描述方式示例如下：

```
## 描述了域名从 www.01.example.com 到 www.50.example.com 的 50 台主机 ##
www.[01:50].example.com
## 描述了域名从 db-a.example.com 到 db-f.example.com 的 6 台主机 ##
db-[a:f].example.com
## 描述了 IP 地址从 192.168.22.1 到 192.168.22.16 的 16 台主机 ##
192.169.22.[1:16]
```

5. inventory 中的变量定义

前面提到的行为参数，其实就是具有特殊意义的 Ansible 变量。在 inventory 文件中，我们可以根据需要任意地指定变量名并对其进行赋值操作。总体来说，变量名的指定和赋值可以分为针对被管理节点的主机变量和针对被管理节点组的组变量。示例如下：

```
mysql.example.com color=red
oracle.example.com color=yellow
postgre.example.com color=green

[vagrant]
vagrant1 ansible_host=127.0.0.1 ansible_port=2200
vagrant2 ansible_host=127.0.0.1 ansible_port=2201
vagrant3 ansible_host=127.0.0.1 ansible_port=2202
```

```
[vagrant:vars]
ansible_user=vagrant
ansible_private_key_file=~/.vagrant.d/insecure_private_key

[all:vars]
db_host=10.172.3.6
db_port=5432
db_name=db_demo
db_user=client
db_password=P@ssw0rd
```

注意:
- [all:vars]段中配置的变量是对整个 inventory 文件中所有被管理节点都有效的,也就是说,所有被管理节点都可以使用[all:vars]中的变量信息,[all:vars]类似于软件设计中的全局变量的概念。
- [vagrant:vars]段中的变量仅限于[vargrant]组中的所有被管理节点使用。[vagrant:vars]类似于软件设计中的对象的私有变量的概念。
- mysql.example.com color=red 中的变量 color 仅能提供给被管理节点 mysql.example.com 使用;即使变量名相同,但是不同的被管理节点访问 color 变量时得到的值却大不相同。

3.1.2 动态 inventory (dynamic inventory)

前面我们明确地描述了 inventory 文件的定义方式和注意事项。但是日常使用中被管理节点可能运行在虚拟化平台或公有云平台上,而且数量巨大。例如,VMWare vCenter、OpenStack、公有云服务商(阿里云、腾讯云、华为云等)。或者已经使用配置管理数据库(CMDB)记录了工作环境中的所有主机信息,也是因为数量巨大而不希望手动将这些主机信息添加到 inventory 文件中。因为在手动添加的过程中很难保证这些数据和外部文件的一致性。因此 Ansible 提供了动态 inventory 功能来避免出现手动复制的过程。

不同于静态 inventory 文件,动态 inventory 文件需要先设置文件的可执行标志,当 Ansible 检查到 inventory 文件被标记为可执行状态后,会假设这是一个动态 inventory 脚本,并且尝试执行它,而不是尝试读取它的内容。

因为篇幅有限,此处就不展开讨论动态 inventory 文件,相关介绍可通过 Ansible 官网了解详情:

- Working with dynamic inventory【链接 1】;

- Developing dynamic inventory【链接2】；
- Using VMware dynamic inventory plugin【链接3】。

注意：
如果需要将多个 inventory 文件结合起来使用，甚至需要将静态和动态 inventory 文件任意组合使用，那么只需要将这些 inventory 文件统一放到一个指定的目录中，并且告知 Ansible 使用这个目录即可。Ansible 会自动处理这个目录中的所有 inventory 文件，并合并为一个完整的 inventory 文件。

告知 Ansible 使用 inventory 的方法主要有两种：

（1）修改 ansible.cfg 文件，例如：

```
[defaults]
inventory=~/project/inventoies
```

（2）在命令行中使用 -i 参数，例如：

```
ansible all -i ~/project/invertories/ -m ping
```

3.2 在命令行中执行 Ansible

在了解了主机资产的配置方法之后，我们先把/etc/ansible/hosts 中的内容清空，加入如下信息，表明我们需要使用用户名为 root、密码为 1q2w3e4r 的账号去对 IP 地址为 192.168.41.139 的主机进行操作：

```
192.168.41.139 ansible_ssh_user=root ansible_ssh_pass=1q2w3e4r
```

配置完成后，我们输出第一条命令：

```
ansible all -a "echo HelloWorld"
```

我们期望目标主机返回 HelloWorld 的结果，但实际上 Ansible 的执行却卡在了询问是否记录 SSH 密钥的位置：

```
paramiko: The authenticity of host '192.168.41.139' can't be established.
The ssh-rsa key fingerprint is 75c4e8af474a9ecbae1e32d3c2550306.
Are you sure you want to continue connecting (yes/no)?
```

这是由于 Ansible 检查服务器存放的 SSH-KEY 造成的，因为我们没有添加双机互信，所以 Ansible 就卡在了询问是否记录 SSH 密钥的位置了。我们可以通过修改配置文件的方式来解决这个问题，不要让 Ansible 对 SSH-KEY 进行检查就可以了。修改配置文件/etc/ansible/ansible.cfg，把 host_key_checking = False 这一句的注释去掉，再执行一次上面的命令，就能得到正确的结果了。

```
192.168.41.139 | success | rc=0 >>
HelloWorld
```

3.2.1 指定目标主机

Ansible 命令的基本语法格式如下所示。其中，pattern 参数声明了需要操作的目标主机，module_name 声明了需要使用的模块是哪一个，最后一个参数 arguments 用于传递参数给模块。

```
ansible <pattern > -m <module_name> -a <arguments>
```

指定所有的主机：

```
ansible all -m shell -a "uptime"
```

指定特定的主机组（同时操作 webservers 和 dbservers 组中的主机）：

```
ansible webservers:dbservers -m shell -a "uptime"
```

排除指定的主机组（排除 webservers 组中所有在 phoenix 组中的主机）：

```
ansible webservers:!phoenix -m shell -a "uptime"
```

指定同时存在于两个组中的主机（操作同时存在于 webservers 和 staging 两个组中的主机）：

```
ansible webservers:&staging -m shell -a "uptime"
```

复合条件：

```
ansible webservers:dbservers:&staging:!phoenix -m shell -a "uptime"
```

采用正则表达式指定主机：

```
ansible one*.com:dbservers -m shell -a "uptime"
```

3.2.2 常用命令示例

以下的示例主要是为了让读者对 Ansible 的常用模块有一个大概的了解，读者可以在自己的环境中亲自验证，有一个大致的印象就可以了。

1. 命令执行

```
## 重启主机 ##
ansible all -a "/sbin/reboot" -f 10
## shell 模块 ##
ansible all -m shell -a 'echo $TERM'
## 底层 SSH 模块 ##
ansible all -m raw -a "hostname --fqdn"
```

2. 文件操作

```
## 下发文件：##
ansible all -m copy -a "src=/etc/hosts dest=/tmp/hosts"
## 为文件赋予指定的权限：##
ansible all -m file -a "dest=b.txt mode=600 owner=demo group=demo"
## 创建文件夹：##
ansible all -m file -a "dest=/path/to/c mode=644 owner=mdehaan group=mdehaan state=directory"
## 删除文件：##
ansible all -m file -a "dest=/path/to/c state=absent"
```

3. 包管理模块

```
## 使用 YUM 源进行安装：##
ansible webservers -m yum -a "name=acme state=installed"
## 安装指定版本的包：##
ansible webservers -m yum -a "name=acme-1.5 state=installed"
## 安装最新版本的安装包：##
ansible webservers -m yum -a "name=acme state=latest"
## 卸载安装包：##
ansible webservers -m yum -a "name=acme state=removed"
```

4. 用户管理

```
## 新增用户：##
ansible all -m user -a "name=demo password=1q2w3e4r"
## 删除用户：##
ansible all -m user -a "name=foo state=absent"
```

5. 版本管理

```
## 使用 Git 拉取文件：##
ansible all -m git -a "repo=git://demo/repo.git dest=/srv/myapp version=HEAD"
```

6. 服务管理

```
## 启动服务：##
ansible webservers -m service -a "name=httpd state=started"
## 重启服务：##
ansible webservers -m service -a "name=httpd state=restarted"
## 停止服务：##
ansible webservers -m service -a "name=httpd state=stopped"
```

7. 后台管理

```
## 启动一个执行 360 秒的后台作业：##
ansible all -B 360 -a "/usr/bin/long_running_operation --do-stuff"
## 检查作业状态：##
ansible all -m async_status -a "jid=1311"
## 后台运行 1800 秒，每 60 秒检查一次作业状态：##
ansible all -B 1800 -P 60 -a "/usr/bin/long_running_operation --do-stuff"
```

8. 设备信息查询

```
## 获取设备的信息列表：##
ansible all -m setup
```

3.3 Ansible 常用模块

Ansible 有许多现成的模块可供我们选用，我们可以使用 "ansible-doc –l" 命令查看 Ansible 内置的模块，下面仅对几个常用的模块进行介绍，并对每个模块的使用场景做一些简要的说明。

其余模块可查看 Ansible 文档说明页面。

说明：

从 Ansible 2.10 开始，Ansible 调整了 Ansible 模块、插件的发布方式，引入了 Ansible Collections 的概念，将单一的模块库分类归到了 Collection 中，详细说明和注意事项详见：Ansible Collections Overview。

3.3.1 文件管理模块

1. 文件组装模块——assemble

assemble 模块用于把多份配置文件片段组装成一份配置文件，当我们需要对不同的主机分配不同的配置文件时，可以考虑使用此模块，组装的方式如图 3-1 所示。

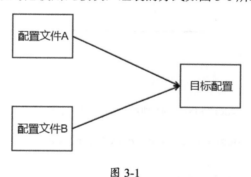

图 3-1

例如，将/root/demo 下的片段文件组装后放到/root/target 目录下。

```
ansible all -m assemble -a 'dest=/root/target src=/root/demo'
```

assemble 模块参数及说明如表 3-2 所示。

表 3-2

参数	是否必选项	默认值	可选值	说明
backup	否	no	yes/no	是否需要备份原始文件
delimiter	否	—	—	配置文件片段之间的分隔符
dest	是	—	—	生成路径
other	否	—	—	文件模块参数
src	是	—	—	片段文件夹路径

2. 文件复制模块——copy

文件复制模块常用于做集中下发的动作,如果被管理节点主机上装有 SELinux,那么我们还需要在目标主机上安装 libselinux-python 模块才能使用 copy 模块。

注意:
SELinux 是一种基于域—类型模型(domain-type)的强制访问控制(MAC)安全系统,它由 NSA 编写并设计成内核模块包含到内核中,相应的某些安全相关的应用也被打了 SELinux 的补丁。最后还有一个相应的安全策略,任何程序对其资源享有完全的控制权。假设某个程序打算把含有潜在重要信息的文件放到/tmp 目录下,那么在 DAC 情况下没人能阻止它。SELinux 提供了比传统的 UNIX 权限更好的访问控制。

使用 copy 模块下发文件,如图 3-2 所示。

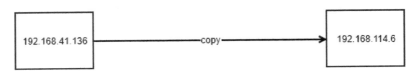

图 3-2

例如,将/root/demo/copydemo.txt 复制到所有主机的/root 目录下。

```
ansible all -m copy -a 'dest=/root src=/root/demo/copydemo.txt'
```

copy 模块参数及说明如表 3-3 所示。

表 3-3

参数	是否必填	默认值	选项	说明
backup	否	no	yes/no	是否备份原始文件
content	否	—	—	当用 content 代替 src 参数时,可以把文档的内容设置为特定的值
dest	是	—	—	文件复制的目的地
force	否	no	yes/no	是否覆盖
others	否	—	—	文件模块参数
src	否	—	—	复制的源文件
validate	否	—	—	复制前是否检验需要复制目的地的路径

3. 文件拉取模块——fetch

fetch 模块和 copy 模块类似,都是对文件进行复制,但 fetch 模块的作用是把被管理节点的文件批量地复制到主机上,可以看作一个文件上传的动作。使用 fetch 模块抓取文件,如图 3-3 所示。

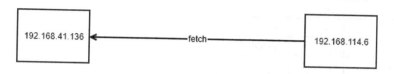

图 3-3

例如将远端机器的/etc/salt/minion 文件收集回主机的/root/demo 目录下。

ansible all -m fetch -a 'dest=/root/demo src=/etc/salt/minion'

fetch 模块参数及说明如表 3-4 所示。

表 3-4

参数	是否必填	默认值	选项	说明
dest	是	—	—	文件存放路径,假如存放路径是/backup,复制的源文件为/etc/profilc,目标主机名是 host,那么文件就会被存放在/backup/host/etc/profilc 下
Fail_on_missing	否	no	yes/no	假如找不到目标文件则标记为失败
flat	否	—	—	用于覆写原有的 dest 存放规则
validate_md5	否	no	yes/no	是否用 md5 进行文件的校验
src	是	—	—	目标文件路径

4. 文件管理模块——file

文件自身有许多属性,如修改文件所属的用户组、文件所属的用户、是否需要删除文件,这些都是我们平常需要使用的功能,而 file 模块就是为完成上述功能而准备的。

例如,删除所有主机下的/root/copydemo.txt 文件。

ansible all -m file -a 'path=/root/copydemo.txt state=absent'

file 模块参数及说明如表 3-5 所示。

表 3-5

参数	是否必填	默认值	选项	说明
force	否	no	yes/no	是否覆盖原有文件
group	否	—	—	文件属于的用户组
mode	否	—	—	文件的读/写权限

续表

参数	是否必填	默认值	选项	说明
owner	否	—	—	文件属于的用户
path	是	—	—	文件路径
recurse	否	no	yes/no	是否递归设置属性
selevel	否	s0	—	SELinux 的级别
serole	否	—	—	SELinux 的角色
setype	否	—	—	SELinux 的类型
seuser	否	—	—	SELinux 的用户
src	否	—	—	文件链接路径
state	否	file	file/link/directory/hard/touch/absent	如果值为 directory，则创建文件夹；如果值为 file，则创建文件；如果值为 link，则创建链接；如果值为 hard，则创建硬链接；如果值为 touch，则创建一份文件；如果值为 absent，则删除文件

5. ini 文件管理模块——ini

ini 文件是十分常见的一种配置文件，Ansible 内置了 ini 配置文件的管理模块，用于对 ini 文件进行配置项的管理。

说明：

ini 文件由节（section）、参数（key=value）和注释组成，格式如下：

```
[section]
key=value ;this is commnet
```

例如，修改/root/demo.ini 配置文件，找到文件中 selection 为 cron 的选项组，修改 crontime 选项，把 cron 的值修改为 10。

```
ansible all -m ini -a 'dest=/root/demo.ini section=cron option=crontime value=10'
```

ini 模块参数及说明如表 3-6 所示。

表 3-6

参数	是否必填	默认值	选项	说明
backup	否	no	yes/no	是否创建备份文件
dest	是	—	—	ini 文件路径
option	否	—	—	ini 文件的键选项

续表

参数	是否必填	默认值	选项	说明
others	否	—	—	文件模块的其他参数
section	是	—	—	选中 ini 的变量名
value	否	—	—	ini 变量的值

3.3.2 命令执行模块

1. 命令执行模块——command

command 模块用于在给定的主机上执行命令。值得注意的是，command 模块执行的命令是获取不到$HOME 这样的环境变量的，一些运算符如 ">""<""|" 在 command 模块上也是不能使用的。

```
ansible all -m command -a 'uptime'
```

command 模块参数及说明如表 3-7 所示。

表 3-7

参数	是否必填	默认值	选项	说明
chdir	否	—	—	执行命令前先进入某个目录
creates	否	—	—	一个文件名，假如文件名已经存在，则不会执行此步骤
executable	否	—	—	改变执行命令所用的 shell
free_form	是	—	—	需要执行的指令
removes	否	—	—	一个文件名，假如不存在该文件，则不会执行此步骤

2. command 模块的增强——shell

前面提到 command 模块是不支持运算符的，也不支持管道这样的操作符。假设我们需要获取 mysql 进程的相关信息，使用 command 模块的操作如下：

```
ansible all -m command -a 'ps -ef|grep mysql'
```

执行上面的命令后，Ansible 会发出不支持操作符的提示，而 shell 模块就支持常见的 shell 语法的相关组合能力，可以视为对 command 模块的功能增强。使用 shell 模块获取 mysql 进程相关信息的方法如下：

```
ansible all -m shell -a 'ps -ef|grep mysql'
```

shell 模块参数及说明表 3-8 所示。

表 3-8

参数	是否必填	默认值	选项	说明
chdir	否	—	—	执行命令前先进入某个目录
creates	否	—	—	一个文件名，假如文件名已经存在，则不会执行此步骤
free_form	是	—	—	需要执行的指令
removes	否	—	—	一个文件名，假如不存在该文件，则不会执行此步骤

3. 脚本执行模块——script

很多时候执行单条命令并不能满足我们的需求，我们需要在目标主机上执行一系列命令。这种情况下我们可以考虑把多条命令写成脚本，然后通过 Ansible 的文件管理模块把脚本下发到目标主机上。接着使用 script 模块执行脚本，得到我们所期望的结果。需要注意的是，执行的脚本是在管理主机上存在的脚本。

例如，我们已经通过 copy 模块把一份巡检脚本下发到了主机上，并且用 file 模块完成了对脚本文件的授权。

inspection.sh 脚本的内容如下：

```bash
#!/bin/bash
phy_cpu=`cat /proc/cpuinfo |grep "physical id"|sort |uniq|wc -l`
logic_cpu_num=`cat /proc/cpuinfo |grep "processor"|wc -l`
cpu_core_num=`cat /proc/cpuinfo |grep "cores"|uniq|awk -F: '{print $2}'`
cpu_freq=`cat /proc/cpuinfo |grep MHz |uniq |awk -F: '{print $2}'`
system_core=`uname -r`
system_version=`head -n 1 /etc/issue`
system_hostname=`hostname`
system_envirement_variables=`env |grep PATH`
mem_free=`grep MemFree /proc/meminfo`
disk_usage=`df -h`
system_uptime=`uptime`
system_load=`cat /proc/loadavg`
system_ip=`ifconfig |grep "inet addr"|grep -v "127.0.0.1" |awk -F: '{print $2}' |awk '{print $1}'`
mem_info=`/usr/sbin/dmidecode | grep -A 16 "Memory Device" | grep -E "Size|Locator" | grep -v Bank`
mem_total=`grep MemTotal /proc/meminfo`
```

```
day01='date +%Y'
day02='date +%m'
day03='date +%d'

path=inspection.txt
echo -e ""> $path
echo -e    $day01 年$day02 月$day03 系统巡检报告 >> $path
echo -e 服务器 IP 地址: "\t"$system_ip >> $path
echo -e 主机名: "\t"$system_hostname >> $path
echo -e 系统内核: "\t" $system_core >> $path
echo -e 操作系统版本:"\t" $system_version >> $path
echo -e 磁盘使用情况: "\t""\t" $disk_usage >> $path
echo -e CPU 核数: "\t" $cpu_core_num >> $path
echo -e 物理 CPU 个数: "\t" $phy_cpu >> $path
echo -e 逻辑 CPU 个数: "\t" $logic_cpu_num >> $path
echo -e CPU 的主频:"\t" $cpu_freq >> $path
echo -e 系统环境变量:"\t" $system_envirement_var >> $path
echo -e 内存简要信息: "\t" $mem_info >> $path
echo -e 内存总大小: "\t" $mem_total >> $path
echo -e 内存空闲: "\t" $mem_free >> $path
echo -e 时间/系统运行时间/当前登录用户/系统过去 1 分钟/5 分钟/15 分钟内平均负载/"\t"
$system_uptime >> $path
echo -e  1 分钟/5 分钟/15 分钟平均负载/在采样时刻，运行任务的数目/系统活跃任务的个数/最大的
pid 值线程/ "\t" $system_load >> $path
```

可以看到整份脚本的命令数量是比较多的，采用 shell 模块或者 command 模块都无法很好地完成这个巡检任务。这时我们可以用 script 模块完成主机的批量巡检。

```
ansible all -M script -a '/root/demo/inspection.sh'
```

script 模块参数及说明如表 3-9 所示。

表 3-9

参数	是否必填	默认值	选项	说明
free_form	是	—	—	需要执行的脚本

4. SSH 命令执行模块——raw

Ansible 虽然不需要安装客户端，但是内置的模块大多需要客户端上有 Python 环境或者具

备某些 Python 扩展才能够执行。假设我们管理的设备上没有 Python 环境，那么 Ansible 的很多模块都用不了，但是又想执行一些简单的命令，怎么办呢？这时我们就可以使用 raw 模块了，这个模块是直接通过 SSH 的方式而不是通过 Python 的方式对目标主机进行操作的。

例如，我们可以通过 raw 模块直接执行一些简单的命令。

```
ansible all -m raw -a 'ip a'
```

raw 模块的参数及说明如表 3-10 所示。

表 3-10

参数	是否必填	默认值	选项	说明
executable	否	—	—	改变执行命令所用的 shell
free_form	是	—	—	需要执行的指令

3.3.3 网络相关模块

1. 下载模块——get_url

get_url 模块用于下载网络上的文件。

例如，下载电子工业出版社的首页。

```
ansible all -m get_url -a 'dest=/root url=http://www.phei.com.cn'
```

get_url 模块参数及说明如表 3-11 所示。

表 3-11

参数	是否必填	默认值	选项	说明
dest	是	—	—	文件下载路径
force	否	no	yes/no	是否覆盖
others	否	—	—	文件模块的其他参数
sha256sum	否	—	—	是否采用 SHA-256 校验和
url	是	—	—	下载文件的目标路径
use_proxy	否	no	yes/no	是否使用代理

2. Web 请求模块——uri

uri 模块主要用于发送 HTTP 协议，通过使用 uri 模块，我们可以让目标主机向指定的网站发送如 Get、Post 这样的 HTTP 请求，并且能得到返回的状态码。

uri 模块参数及其说明如表 3-12 所示。

表 3-12

参数	是否必填	默认值	选项	说明
HEADER_	否	—	—	HTTP 头
Body	否	—	—	HTTP 消息体
Creates	否	—	—	文件名称
Dest	否	—	—	文件下载路径
follow_redirects	否	no	yes/no	uri 模块是否应该遵循所有的重定向
force_basic_auth	否	no	yes/no	强制在发送请求前发送身份验证
method	否	GET	GET/POST/PUT/HEAD/DELETE/OPTIONS	HTTP 方法
others	否	—	—	文件模块参数
password	否	—	—	密码
removes	否	—	—	需要删除的文件名称
return_content	否	no	yes/no	返回内容
status_code	否	200	—	状态码
timeout	否	30	—	超时限制
url	是	—	—	URL 地址
user	否	—	—	用户名

3.3.4 代码管理模块

1. Git

当我们需要将文件集中下发的时候，除了可以用 copy 这样的方式，其实还可以用源码管理模块来实现。

说明：
Git 是 Linus Torvalds 为了帮助管理 Linux 内核而开发的一个开放源码的版本控制系统。Torvalds 开始着手开发 Git 是为了作为一种过渡方案来替代 BitKeeper，后者之前一直是 Linux 内核开发人员在全球使用的主要源代码工具。开源社区中的有些人觉得 BitKeeper 的许可证并不适合开源社区的工作，因此 Torvalds 决定着手研究许可证更为灵活的版本控制系统。

在了解 Git 模块所提供的功能之前，我们先快速地了解一下 Git 这款工具。它与传统的集中式版本管理工具不一样,它有本地仓库这个概念,Git 版本管理模式如图 3-4 所示。

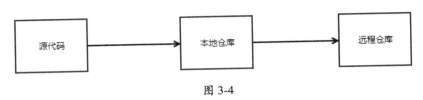

图 3-4

每当我们提交资料的时候,都会先提交到本地仓库,再提交到远端仓库。正因为有了本地仓库这一概念,使得 Git 可以脱离服务器进行版本的控制,等到需要提交或者网络能连接上服务器的时候,再在本地仓库执行一次批量提交的操作。提交到本地仓库的动作叫作 commit,从本地仓库提交到服务器上的动作叫作 push,而从服务器上拉取最新版本的动作叫作 pull,Git 工作流如图 3-5 所示。

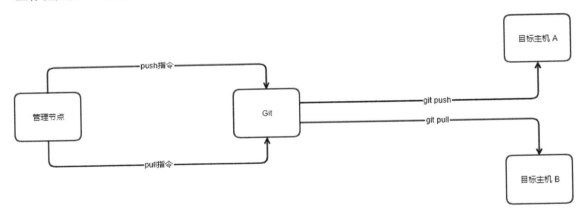

图 3-5

首先,我们在管理节点上把需要下发的文件 "commit" 到本地,然后通过 push 操作将文件传送到远端的 Git,接着通过 Git 模块命令目标主机执行 pull 命令。通过这样一系列的操作,完成文件从管理节点到目标主机的传输过程。

Git 模块参数及说明如表 3-13 所示。

表 3-13

参数	是否必填	默认值	选项	说明
depth	否	—	—	创建一个浅克隆历史阶段到指定的版本
dest	是	—	—	代码克隆的位置
executable	否	—	—	Git 的可执行路径
force	否	no	yes/no	是否强制更新

续表

参数	是否必填	默认值	选项	说明
remote	否	origin	—	远端的版本
repo	是	—	—	仓库地址
update	否	no	yes/no	是否更新远端仓库
version	否	HEAD	—	迁出的版本

3.3.5 包管理模块

Ansible 支持多种包管理模块，由于篇幅有限，此处仅以 APT 和 YUM 为例。

1. APT

Advanced Packaging Tool（APT）是 Linux 下的一款安装包管理工具。通过使用 APT，我们可以非常容易地完成软件的安装。

例如，使用 APT 模块安装 JDK。

```
ansible all -m apt -a 'name=openjdk-8-jdk state=latest install_recommends=no'
```

APT 模块参数及说明如表 3-14 所示。

表 3-14

参数	是否必填	默认值	选项	说明
cache_valid_time	否	—	—	校验缓存是否过期
default_release	否	—	—	等同于 apt –t
force	否	no	yes/no	是否强制更新或删除
install_recommends	否	true	yes/no	等同于 apt –no-install-recommends
pkg	否	—	—	包名
purge	否	—	yes/no	卸载后是否清除配置文件
state	否	present	latest/absent/present	更新、删除、安装包
update_cache	否	—	—	安装前是否更新缓存
upgrade	否	yes	yes/safe/full/dist	safe 等同于 safe-upgrade；full 等同于 full-upgrade；dist 等同于 dist-upgrade

2. YUM

YUM（全称为 Yellow dog Updater，Modified）是一个在运行在 Fedora、Red Hat 和 CentOS

中的 shell 前端软件包管理器。使用 YUM 管理 RPM 包，能够从指定的服务器自动下载 RPM 包并安装，可以自动处理依赖性关系，并且一次性安装所有依赖的软件包，无须烦琐地一次次下载、安装。

例如，使用 YUM 模块安装最新的 httpd。

```
ansible all -m yum -a 'name=httpd state=latest'
```

YUM 模块参数及说明如表 3-15 所示。

表 3-15

参数	是否必填	默认值	选项	说明
conf_file	否	—	—	YUM 配置文件
disable_gpg_check	否	no	yes/no	是否开启 GPG 检查
disablerepo	否	—	—	禁用的仓库
enablerepo	否	—	—	启用的仓库
list	否	—	—	非幂等性命令
name	是	—	—	包名
state	否	present	present/latest/absent	安装、更新、卸载操作

3.3.6 系统管理模块

Ansible 提供了丰富的系统管理模块，本节我们介绍在工作中使用频率比较高的部分管理模块。想要进一步了解更多系统管理模块的功能信息，可以查询 Ansible 官方文档：

- Ansible.Builtin【链接 4】；
- Ansible.Netcommon【链接 5】；
- Ansible.Posix【链接 6】；
- Ansible.Windows【链接 7】。

1. 计划任务管理模块——cron 模块

在 Linux 系统中，推荐使用 crontab 实现定时计划任务，按照 crontab 所需要的格式配置相关参数，系统就会按照配置自动对作业进行定期调度。

打开/etc/crontab 文件，每一行都代表一项任务，每个字段代表一项设置，它有七个字段（如果不指定执行计划任务所属用户，则系统自动默认用户为 root），前五个字段是时间设定段，第六个字段是要执行的命令段，格式如下：

```
# Example of job definition:
# .---------------- minute (0 - 59)
# |  .------------- hour (0 - 23)
# |  |  .---------- day of month (1 - 31)
# |  |  |  .------- month (1 - 12) OR jan,feb,mar,apr ...
# |  |  |  |  .---- day of week (0 - 6) (Sunday=0 or 7) OR sun,mon,tue,wed,thu,fri,sat
# |  |  |  |  |
# *  *  *  *  *  user-name  command to be executed
```

- minute 表示分钟，可以是从 0 到 59 之间的任何整数；
- hour 表示小时，可以是从 0 到 23 之间的任何整数；
- day 表示日期，可以是从 1 到 31 之间的任何整数；
- month 表示月份，可以是从 1 到 12 之间的任何整数；
- week 表示星期几，可以是从 0 到 7 之间的任何整数，这里的 0 或 7 代表星期日；
- command 是要执行的命令，既可以是系统命令，也可以是自己编写的脚本文件。

例如，我们需要每分钟执行一次命令，那么可以在 crontab 中加入如下配置：

* * * * * command

当管理的主机数目比较多的时候，人工到每台主机上执行 crontab 的配置操作显然并不高效，这时我们就可以使用 Ansible 的 cron 模块了。

例如，每天 8 点进行 MySQL 数据库的备份操作。

```
ansible all -m cron -a 'name=demo hour=8 job= mysqldump -uroot -pxxxx demo>demo.sql'
```

cron 模块参数及说明如表 3-16 所示。

表 3-16

参数	是否必填	默认值	选项	说明
backup	否	—	—	修改前是否备份
cron_file	否	—	—	cron 定义文件，如果指定了则读这份文件，不会读用户的 cron.d 文件
day	否	*	—	天
hour	否	*	—	小时

续表

参数	是否必填	默认值	选项	说明
minute	否	*	—	分钟
month	否	*	—	月
weekday	否	*	—	星期几
job	否	—	—	执行的命令
name	否	—	—	对作业的描述
reboot	否	no	yes/no	重启后是否需要执行
special_time	否	—	reboot/yearly/annually/monthly/weekly/daily/hourly	特定的执行时间
state	否	present	present/absent	启用或停用作业
user	否	root	—	执行作业的用户

2. 用户组管理模块——group

使用 group 模块可以对主机进行批量的用户组添加或者删除操作。

例如，为主机批量添加 Zabbix 用户组。

```
ansible all -m group -a 'name=zabbix state=present'
```

group 模块参数及说明如表 3-17 所示。

表 3-17

参数	是否必填	默认值	选项	说明
gid	否	—	—	用户组的 GID
name	是	—	—	用户组的名字
state	否	present	present/absent	新增/删除
system	否	no	yes/no	是否为系统组

3. 用户管理模块——user

user 模块可以用于对目标主机进行批量的用户管理操作。

例如，移除所有主机上的 Zabbix 用户。

```
ansible all -m group -a 'name=zabbix state=absent remove=yes'
```

user 模块参数及相应说明如表 3-18 所示。

表 3-18

参数	是否必填	默认值	选项	说明
append	否	—	—	增加到组
comment	否	—	—	用户账户的描述
createhome	否	no	yes/no	是否创建 home 目录
force	否	no	yes/no	是否强制操作
generate_ssh_key	否	no	yes/no	是否生成 SSH 密钥
group	否	—	—	用户组
groups	否	—	—	以逗号分隔的用户组
home	否	—	—	home 目录
login_class	否	—	—	可以设置用户登录类 FreeBSD、OpenBSD 和 NetBSD 系统
name	是	—	—	用户名
non_unique	否	no	yes/no	相当于 useradd –u
password	否	—	—	密码
remove	否	no	yes/no	相当于 userdel –remove
shell	否	—	—	该用户的 shell
ssh_key_bits	否	2048	—	密钥的位数
ssh_key+comment	否	ansible-generated	—	密钥的说明
ssh_key_file	否	$HOME/.ssh/id_rsa	—	密钥的文件名
ssh_key_passphrase	否	—	—	SSH 密钥的密码
ssh_key+type	否	rsa	—	SSH 密钥的类型
state	否	present	present/absent	新增/删除
system	否	no	yes/no	设置为系统账号
uid	否	—	—	用户的 UID
update_password	否	always	always/on_create	是否需要更新密码

4. 服务管理模块——service

service 模块可以帮助我们批量对服务进行操作。

例如，批量重启 httpd 服务。

```
ansible all -m group -a 'name=httpd state=restarted'
```

service 模块参数及相关参数如表 3-19 所示。

表 3-19

参数	是否必填	默认值	选项	说明
arguments	否	—	—	参数
enabled	否	—	yes/no	开机自启动
name	是	—	—	服务名称
pattern	否	—	—	如果服务没响应，则可查看是否具有指定参数的进程，有则认为服务已经启动
runlevel	否	default	—	OpenRC init 脚本
sleep	否	—	—	如果服务被重新启动，则睡眠很多秒后再执行停止和启动命令
state	否	—	started/stopped/restarted/reloaded	服务的状态

5. 系统信息模块——setup

setup 模块可以获取主机的许多信息，比如这台主机的 IP 地址是多少，主机有哪些环境变量，它是承载在什么样的虚拟化平台之上的。我们在后续编写 Playbook 的时候就会经常使用 setup 模块查询主机的信息。

例如，获取主机配置的信息：

```
ansible all -m setup
```

setup 模块参数及说明如表 3-20 所示。

表 3-20

参数	是否必填	默认值	选项	说明
fact_path	否	/etc/ansible/facts.d	—	fact 的路径
filter	否	*	—	过滤器

3.3.7 文档动态渲染与配置模块

这里的动态渲染是指在工作中，根据主机信息的变化对应地调整在这台主机上部署的应用系统的部分配置信息，如主机的 IP 地址或者主机名等，这种情况多发生在集群化部署或主机信息收集等场景。针对这样的应用场景，Ansible 推荐使用 template 模块对配置文件或信息采集文件进行动态渲染。

template 模块参数及说明如表 3-21 所示。

表 3-21

参数	是否必填	默认值	选项	说明
src	是	—	—	模板文件的源路径,可以是相对路径或绝对路径。该文件必须使用 UTF-8 或兼容格式,但 output_encoding 可以用于控制输出模板文件的编码
dest	是	—	—	将模板文件传输到远程计算机上并渲染后模板文件存放的位置,需要使用绝对路径
backup	否	no	yes/no	创建一个包含时间戳的备份文件,以便在出现故障时进行恢复
force	否	no	yes/no	如果目标文件存在,则确定是否对目标文件进行覆盖 yes——如果目标文件存在,则覆盖远端的目标文件 no——仅在目标文件不存在的情况下,才传输模板文件到远端
mode	否	—	—	渲染后的文件或目录应具有的权限。注意:如果使用八进制数表示权限,则需要在传统描述方式之前添加一个前导 0。例如,644→0644。这是因为 YAML 解释器需要用 0 作为前导值才会被识别为八进制数

续表

参数	是否必填	默认值	选项	说明
group	否	任务执行使用的账号对应的默认用户组	—	指定该文件的所属组
owner	否	任务执行使用的账号对应的默认用户	—	指定该文件的所有者

3.4 自动化作业任务的实现——Ansible Playbook

Playbook 是 Ansible 实现自动化任务的重要组成部分，采用 YAML 语言定义。与前面介绍的 CLI 命令行使用单一模块执行单一任务不同，Playbook 主要通过 YAML 定义，将多个单一的模块功能串联起来，形成完整的自动化批处理能力。通过使用 Playbook，可以将 Ansible 的自动化作业场景推广到更广阔的任务场景中。

3.4.1 Playbook 示例

下面是使用 Playbook 安装 Nginx 服务器的示例：

```yaml
---
- name: playbook demo
  hosts: all
  gather_facts: false
  remote_user: remote_user
  become: true
  become_user: root
  become_method: su
  become_password: P@ssw0rd
  tasks:
    - name: Install / Update nginx server
      yum:
        - pkg: nginx
          state: latest
    - name: Config nginx server
      template:
        - src: webserver.conf.j2
```

```
    dest: /etc/nginx/conf.d/webserver.conf
- name: start nignx sservice
    - service:
      name: nginx
      enabled: yes
      state: started
```

3.4.2 常用的 Playbook 结构

根据任务需要，Playbook 可以由四大部分任意组合构成，分别为：

- 目标定义块（target section）——定义简要执行 Playbook 的被管理节点组及附属信息；
- 变量定义块（variable section）——定义 Playbook 运行时需要使用的变量；
- 任务定义块（task section）——定义将要在被管理节点上执行的任务列表；
- 触发器定义块（handler section）——定义 task 执行完成后需要调用的任务。

1. 目标定义块（target section）

target section 主要用于存放 Playbook 所属任务的目标及其相关附属信息，包括但不限于表 3-22 中的参数。

表 3-22

参数	是否必填	默认值	选项	说明
hosts	是	—	—	执行本次任务的被控节点所属组，可以使用前面提到的集合方式指定被控制节点集合
remote_user	否	root	—	登录到被控制节点所使用的账号
become	否	no	yes / no	设置为 yes，即告知 Ansible 登录被控制节点后，需要进行提权操作
become_mothod	否	sudo	sudo su doas pbrun pfexec dzdo ksu machinectl	提权操作使用的方法
become_user	否	—	—	提权操作的目标账号，需要配合命令行参数 --ask_become_password 或密钥变量文件使用

续表

参数	是否必填	默认值	选项	说明
port	否	22	—	连接被控制节点所需要的端口
gather_facts	否	true	true / false	收集被控节点的 fact 信息
forks	否	5	—	Ansible 中一个 SSH 连接为一个分支，默认 Ansible 会同时创建 5 个分支。分支数量越多，消耗的 CPU 和内存资源会相应增加。可借鉴 Ansbile Tower 的建议按以下公式计算 Ansible 控制节点的最大分支数量：Ansible 每个分支预计需要消耗 100MB 内存，所以内存与分支的需求关系为：（内存总量-2048）/100。例如，Ansible 控制节点实际物理内存为 8GB，最大分支数=（8192-2048）/100≈61。Ansible 每 4 个分支预计需要消耗 1 个 CPU 核心（1 core），所以 CPU 与分支的需求关系为：(CPU 总核心数-1)×4。例如，CPU 总核心数为 4，预留 1 核心保证操作系统的正常运行，最大分支数=(4-1)×7=21
connection	否	ssh	kubectl wrm paramikossh local vmwaretools ssh oc docker buildah ……	与被控制节点建立远程通信连接的方式。Ansible 目前支持 26 种远程连接方式，因篇幅有限，文中仅列出了常用的集中远程连接方式。如有兴趣进一步了解 Ansible 支持的连接方式，可以在部署了 Ansible 系统的设备中使用以下命令进一步了解详情：ansible-doc -t connection -l

2. 任务定义块（task section）

一个完整的 Playbook 对任务内容的定义应以关键字 tasks 为区块的开头，区块中可调用系统自带的 Ansible 模块或自定义的 Ansible 模块。示例如下：

```
---
- name: install and start apache
  gather_facts: false
  hosts: web
  tasks:
```

```
= name: install apache packages
  yum:
    name: httpd
    state: latest
- name: start apache service
  service:
    name: httpe
    state: started
    enabled: yes
```

3. 触发器定义块（handler section）

handler 是 Ansible 提供的一种条件控制机制实现方式，但它并不是唯一的条件控制机制，它类似于程序设计中触发器或者回调函数的实现机制，通常需要和关键字 notify 配套使用，notify 在 task 中声明，以确保条件触发的时候 Ansible 会调用 handler section 中对应的任务内容。

以下展示了 handler section 的使用方式：

```
---
- name: Verify apache installation
  hosts: webservers
  vars:
    http_port: 80
    max_clients: 200
  remote_user: root
  tasks:
  - name: Ensure apache is at the latest version
    ansible.builtin.yum:
      name: httpd
      state: latest

  - name: Write the apache config file
    ansible.builtin.template:
      src: /srv/httpd.j2
      dest: /etc/httpd.conf
    notify:
    - Restart apache

  - name: Ensure apache is running
    ansible.builtin.service:
```

```
    name: httpd
    state: started

handlers:
  - name: Restart apache
    ansible.builtin.service:
      name: httpd
      state: restarted
```

注意：

按照官方文档的描述，handler 适用于在系统上进行更改的场合。结合示例分析，主要用途是重启服务或重启机器。

默认情况下 handler 只会在所有 task 执行完成后才会被执行，哪怕被通知了多次，也只会被调用一次。例如，在 task 中多次修改了 Apache 的配置文件并通知重启 Apache 服务，此时 Ansible 仅会重启 Apache 一次，以避免不必要的重复启动。

如果需要将触发器执行环节调整为 task 任务执行完成前执行 handler 中的任务，则可使用 meta 模块配置示例如下：

```
tasks:
  - name: Write the apache config file
    ansible.builtin.template:
      src: /srv/httpd.j2
      dest: /etc/httpd.conf
    notify:
      - Restart apache

  - name: Flush Handlers
    meta: flush_handlers

  - name: Some other tasks
    ……

handler:
  - name: Restart apache
    ansible.builtin.service:
```

```
      name: httpd
      state: restarted
```

在一个完整的 Playbook 中，可能需要根据不同的任务触发不同的任务内容，这时可以使用关键字 listen 为触发器指定一个主题，task 可以按主题名进行通知，例如：

```
handlers:
  - name: Restart memcached
    ansible.builtin.service:
      name: memcached
      state: restarted
    listen: "restart memcache services"

  - name: Restart apache
    ansible.builtin.service:
      name: apache
      state: restarted
    listen: "restart web services"

tasks:
  - name: Restart everything
    ansible.builtin.command: echo "this task will restart the web services"
    notify: "restart web services"
```

这里需要注意的是，触发程序始终按照 task 触发定义的先后顺序运行，而不是按照触发器定义块中定义的先后顺序；对于使用 listen 的处理流程也是如此。

更多关于触发器的详细说明可参考 Ansible 官方文档：Handlers: running operations on change。

3.4.3　变量的使用

我们仅剩下变量定义块（variable section）没有进行说明，由于变量定义块部分涉及 Playbook 中的变量定义和使用部分，为了便于读者理解，需要整体介绍 Ansible Playbook 中变量的使用，所以我们将变量定义块（variable section）纳入本节进行讲解。

在实际工作中，很多刚上手编写 Ansible Playbook 的用户都曾问过笔者一个类似的问题，Playbook 中保存下来的变量会不会因为新的被管理节点的数据改变而被刷新了？

首先，必须要建立的一个概念就是 Ansible 使用变量来管理不同系统之间的差异。也就是说，针对不同的被管理节点，即使指定的变量名相同，获得的结果也可能存在差异，所以变量和被管理节点的关系是一一对应的。其次，Ansible 对变量的定义要求非常灵活，变量可以被定义在 Playbook、inventory、可重复使用的文件、角色 roles 或者命令行中，甚至可以将 Playbook 中一个或多个模块返回值通过关机键字 register 来创建一个新变量。

1. 有效的变量名

Ansible 的变量名只能包含字母、数字和下画线，且变量名不能以数字开头。Python 关键字和 Playbook 关键字都不能作为有效的变量名。另外，Ansible 允许使用下画线作为变量名的开头，但是在 Ansible 中以下画线开头的变量和其他变量完全是相同的。不要因为在部分编程语言中，以下画线开头的变量多为私有变量，就错误地将 Ansible 中变量的概念混为一谈。

2. 定义变量

- 在 inventory 文件中定义变量

  详见 3.1.1 节。

- 在 Playbook 中定义变量

  在 Playbook 中定义变量，一般都会将变量放置在变量定义块（variable section）中，使用关键字 vars 标识列出变量定义块（variable section）所在的位置，例如：

```
---
- hosts: web
  gather_facts: false
  vars:
    https_port: 443
    ssh_port: 60022
  tasks:
    ......
```

- 在可重复使用的变量文件或角色 roles 中定义变量

  为了避免敏感信息泄露，可以将相关的敏感信息定义到一个可重复使用的变量文件中，从而保证敏感信息和 Playbook 的分离。通过这种方式可以保证分享 Playbook 时不会导致敏感信息的泄露。

  Playbook 如下：

```
---
- hosts: team01
  remote_user: access
```

```
          become: true
          become_user: root
          become_method: su
          vars_files:
            - ./vars/external_vars.yml
          tasks:
            ……
```

external_vars.yml 如下：

```
---
become_password: P@ssw0rd
other_var: other_value
```

- 在命令行中定义变量

    在命令行状态下，可以使用--extra-vars 或-e 参数传递变量。使用 Ansible 命令行传递变量时，可接受 JSON 字符串或 key=value 键值对。

    key=value 键值对：

    ```
    ansible-playbook release.yml --extra-vars "version=1.23.45 other_variable=foo"
    ```

> **注意：**
> 该方法传递的值将被统一解释为字符串类型数据。如果需要传递非字符串数据（例如：布尔值、整数、浮点数、列表等），则建议使用 JSON 格式。

    JSON 字符串格式：

    ```
    ansible-playbook release.yml -e
    '{"version":"1.23.45","other_variable":"foo"}'
    ```

3. 在 Playbook 中注册一个变量

在任务支持过程中，经常需要将某个 task 的输出结果作为新一个 task 执行的判定条件或者参数，这时就需要我们使用关键字 register 将 task 输出结果创建为一个变量，便于在后续任务中使用。

```
---
- hosts: web
  gather_facts: false
```

```
tasks:
  - name: Execute commond
    shell: echo 'test_'`< /dev/urandom tr -dc
'1234567890!@#$%^qwertQWERTasdfgASDFGzxcvbZXCVB' | head -c 8`'!20'
    register: result

  - debug:
      msg: result

  - debug: msg="{{ resutl.stdout }}"
```

注意：
（1）已注册的变量储存在内存中与服务器信息相对应，仅对本次任务中剩余 task 有效。无论 task 执行结果是失败或者被跳过，Ansible 都会自动注册一个该变量，只是该变量的状态是失败或者跳过。除非这个 task 是基于标签 tags 的跳过。
（2）当在一个循环任务中注册了一个变量后，该注册的变量包含循环中每一个项目的值，且变量中的数据库都将包含一个 results 属性值，这个属性包含任务模块的所有响应列表。

4. 引用变量

简单的变量可以使用 {{ variable_name }} 的方式实现对变量的引用，但是遇到通过关键字 register 注册的变量或者需要应用系统 fact 变量时，它们的返回结果集是嵌套 YAML 或者 JSON 的数据结构，如果要访问这些嵌套结构中的值，则需要用以下方式进行访问：

{{ result.stdout }}

或者

{{ ansible_facts["eth0"]["ipv4"]["address"] }}

5. 特殊的系统变量——fact

在工作任务中，我们通常需要尝试探测和获取被管理节点的一些相关信息，Ansible 自动提供了这样的探测功能，与远程系统信息相关的变量在 Ansible 中统称为 fact。在 Ansible 系统中，可以使用 ansbile_facts 变量访问这些数据。

注意：
默认情况下可以使用 ansible_前缀将一些 Ansible fact 作为顶级变量进行访问，例如：
{{ ansible_all_ipv4_addresses[0] }}

由于远端被管理节点的差异，Ansible fact 包含大量的可变数据，如果需要参考了解 Ansible fact，则可查询 Ansible 官方文档：Discovering variables: facts and magic variables。

注意：
在 Playbook 中如未特别说明，会默认收集远端被管理节点信息。这在针对不需要使用 fact 的任务中会带来较长的信息收集等待时间，为了省去这个信息收集的过程，可以在目标定义块（target section）中部分显式地声明 gather_facts: false，即可禁用 fact 的采集过程。

6. 变量调用的先后顺序

Ansible 的变量可以在很多地方被声明，如果不小心在配置过程中同时声明了一个同名变量，则会导致同名变量的覆盖问题。Ansible 官方文档罗列了 22 种变量的覆盖场景，场景编号越大，被覆盖的优先级越小。为了节省篇幅就不再重复罗列，详情可查询官方文档：Understanding variable precedence。

3.4.4 条件语句

对服务器进行批量操作时，我们需要根据不同的情况进行不同的操作。这时就需要用到 PlayBook 的条件判断功能。例如，当运行的前置命令成功了，才执行后续的命令。或者是针对不同的操作系统、不同的 IP 地址段执行不同的操作。为了达到这些目的，需要具备一些条件判断的功能，而 Ansible 提供了关键字 when 来实现非常强大的条件功能供我们使用。

例如，需要关闭所有操作系统为 CentOS 的被管理节点：

```
---
- tasks:
  - name: "shutdown CentOS flavored systems"
    command: /sbin/shutdown -t now
    when: {{ ansbile_os_family }} == "CentOS"
```

when 语句还可以和 Jinja2 的过滤器配合使用，例如：

```
---
- tasks:
  - command: /bin/false
    register: result
    ignore_errors: true
  - shell: do_something.sh
    when: {{ result | failed }}
```

when 语句还可以判断变量是否被定义，例如：

```
---
- tasks:
    - shell: echo "I've got '{{ foo }}' and I am not afraid to use it"
      when: foo is defind

    - debug: msg="Bailing out: this play requires 'bar'"
      when: bar is defind
```

3.4.5 循环控制

Ansible Playbook 中可使用多种循环控制的写法。

1. 常见的写法

```
- name: add several users
  user: name={{ item }} state=present groups=wheel
  with_items:
     - testuser1
     - testuser2
```

2. 用 Hash 表做循环变量

```
- name: add several users
  user: name={{ item.name }} state=present groups={{ item.groups }}
  with_items:
     - { name: 'testuser1', groups: 'wheel' }
     - { name: 'testuser2', groups: 'root' }
```

3. 把文件名称作为变量循环

```
---
- hosts: all
  tasks:
     - file: dest=/etc/fooapp state=directory
     - copy: src={{ item }} dest=/etc/fooapp/ owner=root mode=600
```

```
with_fileglob:
  - /playbooks/files/fooapp/*
```

4. 复合变量循环

假如变量的格式如下：

```
---
- alpha: [ 'a', 'b', 'c', 'd' ]
  numbers:  [ 1, 2, 3, 4 ]
```

执行循环逻辑的时候期望传入的变量为(a,1) (b,2)这样的组合，则可以采用如下的方法：

```
---
- tasks:
    - debug: msg="{{ item.0 }} and {{ item.1 }}"
      with_together:
        - alpha
        - numbers
```

5. 步进变量循环

循环的变量为整型，如 i=0;i<100;i++这种情况：

```
---
- hosts: all
  tasks:
    - group: name=evens state=present
    - group: name=odds state=present

    - user: name={{ item }} state=present groups=evens
      with_sequence: start=0 end=32 format=testuser%02x

    - file: dest=/var/stuff/{{ item }} state=directory
      with_sequence: start=4 end=16 stride=2

    - group: name=group{{ item }} state=present
      with_sequence: count=4
```

6. 随机变量循环

```
- debug: msg={{ item }}
  with_random_choice:
    - "go through the door"
    - "drink from the goblet"
    - "press the red button"
    - "do nothing"
```

7. Do-Until 类型的循环

```
- action: shell /usr/bin/foo
  register: result
  until: result.stdout.find("all systems go") != -1
  retries: 5
  delay: 10
```

3.4.6 include 语法

前面主要介绍了在编写 Ansible Playbook 时常用的语法和相关功能模块，借鉴早期的模块化程序设计思想，将一个完整的任务放在单一的 Playbook 文件中会导致文件长度过大，而且还很难实现代码重复利用的目标。为此，Ansible 提供了多种代码复用的实现方式，关键字 include 就是其中一种简单的实现方式。

简单的 include 用法：

```
---
- tasks:
    - include: tasks/foo.yml
```

引用配置文件的同时传入变量：

```
---
- tasks:
    - include: a.yml user=timmy
    - include: b.yml user=alice
    - include: c.yml user=bob
```

3.4.7 Ansible Playbook 的角色 roles

角色 roles 是 Playbook 结构化代码的另一种重要实现方式，通过规范的目录存储结构有效地将任务和被管理节点信息分离，通过一个个独立的 role 目录将不同的 Playbook 任务单一化，从而实现高效的代码复用的目标。

1. roles 的目录结构

Ansible roles 需要特定的目录结构，其中有七个主要的标准目录，每个 role 必须至少包含这些目录之一，用不到的目录可以忽略不设置。例如：

```
[root@gitlab ansible_roles]# tree -a
.
├── hosts
│   └── inventory
├── library
│   └── other_module.py
├── roles
│   ├── export_results
│   │   ├── tasks
│   │   │   └── main.yml
│   │   └── vars
│   │       └── main.yml
│   └── health_check
│       ├── defaults
│       │   └── main.yml
│       ├── files
│       │   └── main.yml
│       ├── handlers
│       │   └── main.yml
│       ├── library
│       │   └── my_module.py
│       ├── meta
│       │   └── main.yml
│       ├── tasks
│       │   └── main.yml
│       ├── templates
```

```
|       |    └── main.yml
|       └── vars
|           └── main.yml
└── site.yml
```

默认情况下，Ansible 将在 roles 中自动查找每个目录中以 main.yml、main.yaml 和 main 命名的文件，并读取其中的内容。上面的目录结构示例中各个目录结构的说明如下：

- site.yml ——roles 的整体编排文件；
- hosts/inventory ——存放被管理节点信息；
- library/other_module.py ——存放用户自定义 Ansible 模块；
- roles/ ——存放 Playbook 中需要用到的各个角色 roles 模块；
    - health_check/ ——角色 roles 模块名，在 Playbook 中需要调用时使用的名称；
        - tasks/main.yml ——角色 roles 执行的主要任务列表；
        - handlers/main.yml ——处理程序，可以在此角色 roles 内部或外部使用；
        - library/my_module.py ——可以在该角色 roles 中使用的 Ansible 自定义模块；
        - defaults/main.yml ——角色 roles 的默认变量；
        - vars/main.yml ——角色 roles 的其他变量；
        - files/main.yml ——角色 roles 部署的文件；
        - templates/main.yml ——角色 roles 部署的模板；
        - meta/main.yml ——角色 roles 的元数据，包括角色依赖性。

2. 调用 roles

在 Playbook 中可以通过以下三种方式实现 roles 的调用：

- 在整个任务编排文件中直接使用关键字 roles 调用角色 roles，这种方式是最典型的角色 roles 调用方式；
- 在 tasks 中使用关键字 include_role 动态调用角色 roles；
- 在 tasks 中使用关键字 import_role 静态态调用角色 roles。

3. 使用 roles 关键字调用角色 roles

```
---
- hosts: webservers
  roles:
```

```
- common
- webservers
```

4. 使用 include_role 关键字调用角色 roles

```
---
- hosts: webservers
  tasks:
    - name: Include the some_role role
      include_role:
        name: some_role
      when: "ansible_facts['os_family'] == 'RedHat'"
```

5. 使用 import_role 关键字调用角色 roles

```
---
- hosts: webservers
  tasks:
    - name: Print a message
      ansible.builtin.debug:
        msg: "before we run our role"

    - name: Import the example role
      import_role:
        name: example

    - name: Print a message
      ansible.builtin.debug:
        msg: "after we ran our role"
```

6. 详细说明

上述示例代码均引用自 Ansible 官方文档，更多关于角色 roles 的介绍，可通过 Ansible 官方文档了解详细信息：Roles。

3.5　密钥管理方案——ansible-vault

通过前面的介绍，是否一直有一种担心——在开发 Playbook 的过程中很容易导致敏感数据的暴露？为了解决这个问题，Ansible 提供了命令行工具 ansible-vault 来对数据文件进行加解密操作，从而更加有效地保护敏感数据。

我们可以使用以下两种方式创建一个新的加密文件：

```
ansible-vault encrypt secrets.yml
ansible-vault create secrets.yml
```

系统将提示输入密码，然后通过系统的环境变量$EDITOR 启动系统所指定的文本编辑器，以便输入需要被加密的信息。如果没有指定环境变量，则默认会启动 vim 编辑器。

在 Playbook 中，可以在变量定义块（variable section）中使用关键字 vars_files 直接引用加密后的 secrets.yml 文件。在执行 Playbook 时，仅需要针对 ansible-playbook 命令添加--ask-vault-pass 或--vault-password-file <file_name>参数，Playbook 即可对 ansible-vault 命令进行自动解密。

常用的 ansible-vault 命令及说明如表 3-23 所示。

表 3-23

命令	说明
ansible-vault encrypt file.yml	加密纯文本文件 file.yml。如果文件不存在则创建一个文件
ansible-vault decrypt file.yml	解密被加密后的文件 file.yml
ansible-vault view file.yml	查看加密文件 file.yml 的内容
ansible-vault create file.yml	创建一个新的加密文件 file.yml
ansible-vault edit file.yml	编辑加密文件 file.yml
ansible-vault rekey file.yml	修改加密文件 file.yml 的密码

3.6 使用 Ansible 的 API

Ansible 除了有直接命令调用、PlayBook 调用的方式，还支持直接通过 API 调用的方式，这种调用方式主要是针对开发者而设计的。

简单调用：

```python
#!/usr/bin/python

import ansible.runner

runner = ansible.runner.Runner(
  module_name='ping',
  module_args='',
  pattern='web*',
```

```
    forks=10
)

datastructure = runner.run()
```

调用成功后，Ansible 会返回 JSON 格式的字符串。

获取所有服务器的启停信息：

```
#!/usr/bin/python

import ansible.runner
import sys

results = ansible.runner.Runner(
  pattern='*', forks=10,
  module_name='command', module_args='/usr/bin/uptime',
).run()

if results is None:
  print "No hosts found"
  sys.exit(1)

for (hostname, result) in results['contacted'].items():
  if not 'failed' in result:
    print "%s >>> %s" % (hostname, result['stdout'])

print "FAILED *******"

for (hostname, result) in results['contacted'].items():
  if 'failed' in result:
    print "%s >>> %s" % (hostname, result['msg'])

print "DOWN *********"

for (hostname, result) in results['dark'].items():
  print "%s >>> %s" % (hostname, result)
```

3.7 Ansible 的优点与缺点

1. Ansible 的优点

Ansible 从诞生之初迭代至今，已经形成了比较完备的生态环境，可使用多种连接方式管理各种操作系统、网络设备和云计算平台。Ansible 具备以下优点：

（1）部署成本极低：不需要对被管理的目标主机安装 Agent 是一件十分惬意的事情。

（2）没有 Agent 更新的问题：正因为 Ansible 是无 Agent 的集中化运维软件，所以它也就没有 Agent 更新的问题，极大地简化了 Agent 维护成本。

（3）学习成本低：Ansible 的操作方式中大量现成的模块减少了我们不少工作量，命令式的操作思路与常规的命令行操作思路十分类似，让使用者非常容易接受。

（4）完备的模块：拥有许多现成的模块，涵盖了许多日常运维所需要的功能。

2. Ansible 的缺点

事情总是有两面性的，目前 Ansible 最大的缺点在于：目标主机需要 Python 解释器；Ansible 之所以会有 raw 这个 SSH 模块，就是为了解决使用 Ansible 操作目标主机时，目标主机缺少相应 Python 模块的问题。在日常接触到的操作系统中，我们运维的 AIX 系统和 Ubuntu 系统默认没有安装 Python 环境，更别说其他 Python 模块了。为了更好地纳管这些操作系统，就需要我们在这几类操作系统中手动部署 Python 环境。

第 4 章 自动化运维

第 3 章介绍 Ansible 时曾提到，Ansible 目前以 SSH 为主，已经广泛支持多种连接方式，可以采用无 Agent 的方式来纳管更多的被管理节点，这为 Ansible 提供了更广泛的适用场景。为了后续叙述方便，本章暂时将自动化运维场景聚焦于 Linux 系统。

4.1 Ansible 在自动化运维中的应用

4.1.1 ansible_fact 缓存

ansible_fact 这个特殊的变量主要用于在内存中保存远程被管理节点中的服务器信息数据。在实际工作中，已知被管理节点信息变化不大的情况下，每次使用 Playbook 执行任务，都要收集一次被管理节点的信息数据就显得低效且无意义。我们可以使用 fact 的缓存设置将这些信息和数据集保存下来，供其他应用程序调用。Ansible 主要支持的 fact 缓存方式如表 4-1 所示。

表 4-1

缓存方式	说明
jsonfile	以 JSON 格式将 fact 数据缓存到文件中
mongodb	将 fact 数据缓存到 MongoDB 中
redis	将 fact 数据缓存到 Redis 中
yaml	以 YAML 格式将 fact 数据缓存到文件中

续表

缓存方式	说明
memory	将 fact 保存在内存中,默认方式
pickle	以 Pickle 格式将 fact 数据缓存到文件中
memcached	将 fact 数据缓存到 memcached DB 中

1. 配置 JSON 缓存

和变量赋值一样，Ansible 支持多种方式设置 fact 缓存功能。这里推荐在 Playbook 同级目录下创建一个 ansible.cfg 类来配置 fact 缓存。JSON 缓存文件的配置信息如下：

```
[defaults]
fact_caching=jsonfile
fact_caching_cannection=~/fact_caching/
```

这里的关键字 fact_caching_connection 需要配置的是存放 fact 缓存文件的路径，既可以是绝对路径，也可以是相对路径。

2. 配置 Redis 缓存

```
[defaults]
fact_caching=redis
fact_caching_connection=localhost:6379
fact_caching_timeout=86400
```

这里的关键字 fact_caching_connection 需要配置的是一个以":"分隔用于连接信息的字符串，格式为：[:][:]。

关键字 fact_caching_timeout 是指缓存数据的失效时间，单位为秒。86400 是系统默认的缓存数据失效时间。

3. 配置 MongoDB 缓存

```
[defaults]
fact_caching=mongodb
fact_caching_connection=mongodb://mongodb0.example.com:27017
```

这里的关键字 fact_caching_connection 需要配置的是一个 MongoDB 数据库连接字符串。

4.1.2 ansible_fact 信息模板

为了便于读者理解后续内容，这里引用 Ansible 官方网站中的一个 ansible_fact 信息描述示例，该示例以 JSON 格式展示 ansible_fact 中一个被控制节点的信息。

注意：根据操作系统的不同，该信息模板也存在差异，建议提前收集相关被控制节点的 ansible_fact 信息并进行比较。详细模板信息格式如下：

```
{
    "ansible_all_ipv4_addresses": [
        "REDACTED IP ADDRESS"
    ],
    "ansible_all_ipv6_addresses": [
        "REDACTED IPV6 ADDRESS"
    ],
    "ansible_apparmor": {
        "status": "disabled"
    },
    // ……
    "ansible_userspace_architecture": "x86_64",
    "ansible_userspace_bits": "64",
    "ansible_virtualization_role": "guest",
    "ansible_virtualization_type": "xen",
    "gather_subset": [
        "all"
    ],
    "module_setup": true
}
```

通过这个示例模板，可以发现 ansible_fact 记录了非常详细的被管理节点信息，甚至包括部分硬件信息、系统配置信息、操作系统类型、版本信息等。因此，我们在很多自动化运维任务中可以直接使用这部分信息。后续的章节中"磁盘挂载点检查"的功能就可以直接使用 ansible_fact 中关于磁盘挂载点的数据信息。

4.1.3 载入 fact

使用缓存的目的在于可以在需要重复使用服务器信息的时候能够迅速访问这些服务器的

fact 信息，重新载入方法如下：

`{{ hostvars['server_a.example.com']['ansible_facts']['os_family'] }}`

4.1.4 set_fact 的使用

在运维任务中，常常希望把一些采集到的数据变量像 fact 一样被保存下来，便于我们可以跨 Playbook 使用。这时可以使用 set_fact 关键字声明哪些变量需要被保存。变量可以直接被写入 fact 缓存，例如：

```
- name: Setting facts so that they will be persisted in the fact cache
  set_fact:
    one_fact: something
    other_fact: "{{ local_var * 2 }}"
    cacheable: yes
```

这里的关键字 cacheable 就是用于告知 Ansible 如果 Playbook 启用了缓存，则将 set_fact 中设置的内存变量转存到指定的持久化缓存中。如果没有这个关键字或者 cacheable: no，则这些变量仅驻留内存。

4.1.5 自定义 module

既然可以在 Playbook 中临时将变量声明为 fact，那么 Ansible 是不是也支持自己编写收集 fact 功能的模块呢？答案是肯定的，fact 的系统信息收集功能其实被划为了 Ansible 自定义 module 中的一个特殊的自定义开发模块。Ansible 默认提供了一个使用 Python 语言开发的 module 模板。在这里就不单独列出 fact module，可以参考官方文档中使用的标准 Ansible 自定义 module 的开发模板。

> **注意：**
> 模板中 DOCUMENTATION、EXAMPLES、RETURN 三个字符串变量是用于定义 module 的相关信息的，为了便于其他人能够使用这个自定义 module，建议按照模板的要求完整地填写和补充模板信息。

更详细的说明可参考 Ansible 官方文档：

- Discovering variables: facts and magic variables【链接 8】；
- Developing plugins【链接 9】。

4.2 挂载点使用情况和邮件通知

前面主要介绍了当 Ansible 标准 module 不能满足运维任务需求时，我们可能用到的一些实现指定运维任务的技术手段。本节以案例的方式介绍如何完善一个特定的运维任务目标。

4.2.1 任务目标

检查被管理节点上各个磁盘挂载点的空间使用情况和 inode 使用情况，允许用户自行指定警戒值，只要超过用户指定的空间使用率或 inode 使用率，通过 CMDB 查询出挂载点的管理员，由 DMZ 区的一台服务器发送报警邮件告知挂载点的管理员。

4.2.2 任务分析

首先，可以直接使用 fact 中的数据检查被管理节点上的各个挂载点信息，ansible_fact 中有一个专门记录挂载点使用情况的 JSON 对象，叫作 ansible_mounts。通过对 ansible_mounts 中数据的简单计算，即可迅速判断各个挂载点的数据信息。然后根据被管理节点的 IP 地址可以在 CMDB 中查询到挂载点管理员的邮箱地址。

其次，DMZ 区的一台服务器也是被管理节点，需要负责发送告警邮件给挂载点管理员。收集数据信息在先，发送邮件在后，如果使用驻留内存的变量，则可能导致控制节点在进行任务变更时无法得到被管理节点数据的清单列表。为了保证数据的准确性，优选使用 fact 对收集到的服务器数据进行缓存。

再次，发送邮件的时候，为了防止告警风暴，将运维工程师所管理的多个挂载点同时发出的告警信息进行告警合并，形成一封邮件通知给运维工程师，是一种实践中推荐的做法。所以需要在发送邮件前将 fact 数据从以服务器为中心整理为以挂载点管理员为中心。

最后，由于 Ansible 系统自带的 mail 模块不支持 when 操作，所以需要单独提供邮件发送功能。

综上所述，我们最少需要编写 3 个自定义 module，分别用于检查挂载点使用情况和查询挂载点管理员；载入 fact 缓存数据并整理成以挂载点管理员为中心的数据目标；以挂载点管理员邮箱地址为依据批量将报警的挂载点信息整理成一个完整的报警邮件，并发送给邮箱管理员。

4.2.3 任务的实现

1. 目录结构

```
.
├── ansible.cfg
├── caching
│   ├── 10.173.245.133
│   ├── 10.174.66.176
│   └── 192.168.1.100
├── demo.html
├── inventory
│   └── host
├── library
│   ├── __init__.py
│   ├── check_mountpoints.py
│   ├── read_data.py
│   └── sendmail.py
└── main.yml
```

这里的配置文件 ansible.cfg 用于启用 jsonfile 缓存和指定自定义 module 所在的路径。内容如下：

```
[defaults]
fact_caching=jsonfile
fact_caching_connection=./caching/
library=./library/
inventory=./inventory/host
```

注意：
如果自定义模块文件存放在 Playbook 所在的 library/ 子目录中，那么默认可以不单独配置。

2. 挂载点检查模块

挂载点模块配置示例：

```python
#!/usr/bin/python
# -*- coding: utf-8 -*-
from __future__ import division
```

```
from ansible.module_utils.basic import AnsibleModule
import traceback
import requests

DOCUMENTATION = """
---
    module: check_mount_point
    short_description: Check system mount point utilization and issue warning messages based on the limits
    description:
        - This module is used to check the utilization of the system's mount point and send warning mail to the
          corresponding system administrator based on the situation of each mount point limit.
    version_added: "1.0"
    options:
        ip_address:
            description:
                - Information about the remote ip address
            default: {{ inventory_host }}
            required: true
            type: str
        hostname:
            description:
                - Information about the remote machine's hostname
            default: {{ ansible_fqdn }}
            required: true
            type: str
        mount_data:
            description:
                - Mount point data, Example: {{ ansible_mounts }}
            default: {{ ansible_mounts }}
            required: true
            type: list
        alter_value:
            description:
                - Alarm thresholds for system mount points managed by application development
```

```
                    required: true
                    default: 90
                    type: int
        author: Sun Jingchong
"""

EXAMPLES = '''
    - name: check software update
      check_mount_point:
        host="{{ inventory_host }}"
        mount_data="{{ ansible_mounts }}"
        alter_value=90
'''

RETURN = '''
# These are examples of possible return values, and in general should use other names
for return values.
hostname:
    description: information about host
    type: str
    returned: always
    sample: "local.localhost" or "demo.example.com"
ip_address:
    description: information about host's ip address
    type: str
    returned: always
    sample: "127.0.0.1" or "192.168.1.2"
result:
    description: Usage information for the mount points.
    type: list
    returned: always
    sample: [
        {
            "mount": "/",
            "inode_available": 10052341,
            "inode_total": 10419200,
            "inode_usage_%": "3.52"
            "size_available": 29577207808,
```

```
            "size_total": 40211361792,
            "size_usage_%": "26.45",
            "manager": "admin@example.com",
            "is_warning": False
        },
        {
            "mount": "/home/oracle",
            "inode_available": 0,
            "inode_total": 0,
            "inode_usage_%": "0",
            "size_available": 0,
            "size_total": 516096,
            "size_usage_%": "100",
            "manager": "system@example.com",
            "is_warning": True
        }
    ]
...


def get_manager(ip, mount_name):
    # 用 requests 向 CMDB 请求查找挂载点及相关负责人的信息
    ......


# Ansible module 主函数
def run_module():
    module = AnsibleModule(
        argument_spec=dict(
            ip_address=dict(required=True, type='str'),
            hostname=dict(required=True, type='str'),
            mount_data=dict(required=True, type='list'),
            system_alter_value=dict(default='90', type='int'),
            application_alter_value=dict(default='90', type='int')
        ),
        supports_check_mode=True
    )

    result = dict(
        changed=False,
```

```python
        inspection_info=None
    )

    try:
        ip_address, hostname, mount_data, system_alter = (module.params["ip_address"],
                                                           module.params['hostname'],
                                                           module.params['mount_data'],
                                                           module.params['alter_value'])

        mount_pointers = []

        for item in mount_data:
            if item['inode_total'] == 0:
                inode_usage = 0
            else:
                inode_usage = round((item["inode_total"] - item["inode_available"]) / item["inode_total"] * 100, 2)

                size_usage = round((item["size_total"] - item["size_available"]) / item["size_total"] * 100, 2)

                manager = get_manager(ip_address, item["mount"])
                is_warning = (inode_usage >= system_alter) or (size_usage >= application_alter)

                data = {}
                data.setdefault("mount", item["mount"])
                data.setdefault("inode_total", item["inode_total"]),
                data.setdefault("inode_available", item["inode_available"]),
                data.setdefault("inode_usage_%", inode_usage)
                data.setdefault("size_total", item["size_total"]),
                data.setdefault("size_available", item["size_available"]),
                data.setdefault("size_usage_%", size_usage)
                data.setdefault("manager", manager)
                data.setdefault("is_warning", is_warning)

                mount_pointers.append(data)
```

```python
        info = {}
        info.setdefault("hostname", hostname)
        info.setdefault("ip_address", ip_address)
        info.setdefault("result", mount_pointers)

        result['changed'] = True
        result["inspection_info"] = info
    except Exception:
        parameter = dict(
            ip_address=module.params["ip_address"],
            hostname=module.params['hostname'],
            mount_data=module.params['mount_data'],
            system_alter=module.params['system_alter_value'],
            application_alter=module.params['application_alter_value']
        )
        result['changed'] = False
        result.setdefault('failed', True)
        result.setdefault('data', parameter)
        result.setdefault('exception', traceback.format_exc())
    finally:
        module.exit_json(**result)


if __name__ == '__main__':
    run_module()
```

3. 读取并整理缓存数据

在执行运维任务之前,我们需要获取数据并整理,然后缓存数据:

```python
#!/usr/bin/python
# -*- coding: utf-8 -*-

import json
from os import listdir, path
from ansible.module_utils.basic import AnsibleModule
import traceback

DOCUMENTATION = """
```

```yaml
---
    module: read_data
    short_description: Load data from caching file.
    description:
        - The result of reading the detected data from the ANSIBle_FACT cache file
    version_added: "1.0"
    options:
        cachefile_path:
            description:
                - Fact cacheing file storage's path, must be an absolute path
            default: ~/ansible.monitor.mountpointer/caching
            required: true
            type: str

    author: Sun Jingchong
"""

EXAMPLES = '''
    - name: Load inspect result from cache file
      local_action: read_data
      register: result
'''


RETURN = '''
# These are examples of possible return values, and in general should use other names for return values.
data:
    description: data information
    type: list
    returned: always
    sample: [
        {
            "manager": "system@example.com",
            "mounts": [
                {
                    "hostname": "gitlab.os3c.cn",
                    "ip_address": "git.os3c.cn",
```

```json
                "inode_available": 3680247,
                "inode_total": 3932160,
                "inode_usage_%": 6.41,
                "is_warning": false,
                "mount": "/",
                "size_available": 49305190400,
                "size_total": 63278391296,
                "size_usage_%": 22.08
            },
            {
                "hostname": "www.os3c.cn",
                "ip_address": "192.168.1.100",
                "inode_available": 3680247,
                "inode_total": 3932160,
                "inode_usage_%": 6.41,
                "is_warning": false,
                "mount": "/",
                "size_available": 49305190400,
                "size_total": 63278391296,
                "size_usage_%": 22.08
            },
            {
                "hostname": "www.os3c.cn",
                "ip_address": "192.168.1.100",
                "inode_available": 3680247,
                "inode_total": 3932160,
                "inode_usage_%": 6.41,
                "is_warning": false,
                "mount": "/var",
                "size_available": 49305190400,
                "size_total": 63278391296,
                "size_usage_%": 22.08
            }
        ]
    },
    {
        "manager": "test@example.com",
        "mounts": [
```

```
            {
                "hostname": "www.os3c.cn",
                "ip_address": "192.168.1.100",
                "inode_available": 3680247,
                "inode_total": 3932160,
                "inode_usage_%": 6.41,
                "is_warning": false,
                "mount": "/home/test",
                "size_available": 49305190400,
                "size_total": 63278391296,
                "size_usage_%": 22.08
            }
        ]
    }
]
...

# 获取文件清单
def getFiles(parent_dir):
    files = []
    if path.isabs(parent_dir):

        for fileName in listdir(parent_dir):
            files.append(path.join(parent_dir, fileName))
    else:
        files = listdir(parent_dir)
    return files


# 载入缓存的 JSON 文件数据
def readJSONFile(filename):
    with open(filename, 'r') as fileHandler:
        return json.load(fileHandler)



# 查找同名节点
def findChildNode(source_data, manager):
    for item in source_data:
        if item.get("manager") == manager:
```

```python
            return item

    return None

# 整理单台服务器的 JSON 数据,将挂载数据统一到管理员名下
def disposalData(source_data):
    data = source_data.get('inspection_info', None)
    if data:
        owners = []
        target_data = list()
        for item in data.get('result'):
            manager = item.get('manager')

            data_node = {}
            data_node.setdefault("hostname", data.get("hostname", None))
            data_node.setdefault("ip_address", data.get("ip_address", None))
            data_node.update(item)
            data_node.pop('manager')

            if manager in owners:
                sub_data = findChildNode(target_data, manager)
                mounts = sub_data.get("mounts")
                mounts.append(data_node)
            else:
                owners.append(manager)

                sub_data = dict()
                sub_data.setdefault('manager', manager)
                sub_data.setdefault("mounts", [data_node])

                target_data.append(sub_data)

        return target_data
    else:
        return None

# 将单台服务器数据合并成完整的数据
def mergeData(source_data, memory_cache):
```

```python
    owner = []
    for item in memory_cache:
        manager = item.get("manager")
        owner.append(manager)

    for item in source_data:
        manager = item.get("manager")
        if manager in owner:
            sub_data = findChildNode(memory_cache, manager)
            mounts = sub_data.get("mounts")
            mounts.extend(item.get("mounts"))
        else:
            memory_cache.append(item)

    return memory_cache

# Ansible module 的主函数
def run_module():
    module = AnsibleModule(
        argument_spec=dict(
            cachefile_path=dict(required=True, type='str')
        ),
        supports_check_mode=True
    )

    result = dict(
        changed=False,
        data=''
    )

    cache_file_path = module.params['cachefile_path']

    # 缓存数据加载过程
    data = {}
    cache = []

    files = getFiles(cache_file_path)
    for item in files:
        # noinspection PyBroadException
```

```python
    try:
        data = readJSONFile(item)
        if data:
            data = disposalData(data)
            if data:
                cache = mergeData(data, cache)
            else:
                data = dict(
                    file=item,
                    description='No have "inspection_info" json object.'
                )
                cache.append(data)
        else:
            data = dict(
                file=item,
                description='Read data failed from: %s' % item
            )
            cache.append(data)
    except Exception:
        result["changed"] = False
        result.setdefault("failed", True)
        result.setdefault("cache_file", item)
        result.setdefault("data", data)
        result.setdefault("exception", traceback.format_exc())
        module.exit_json(**result)

result["changed"] = True
result["data"] = cache

module.exit_json(**result)


if __name__ == '__main__':
    run_module()
```

4. 发送邮件

执行完运维任务之后，给相关负责人发送邮件，方便相关负责人了解本次运维任务的执行结果：

```python
#!/usr/bin/python
# -*- coding: utf-8 -*-

import time
import smtplib
import traceback
from ansible.module_utils.basic import AnsibleModule
from email.mime.text import MIMEText
from email.mime.multipart import MIMEMultipart

htmlTemplate = """

<body>
    <div class="main">
        <div id="advisories">
            <h1>Linux 系统挂载点检查报警通知</h1>
            <table class="table">
                <thead>
                    <tr role="row">
                        <th rowspan=2>服务器名</th>
                        <th rowspan=2>IP 地址</th>
                        <th rowspan=2>挂载点</th>
                        <th colspan=3>inode 检查</th>
                        <th colspan=3>剩余容量检查</th>
                    </tr>
                    <tr>
                        <th>可用节点</th>
                        <th>总节点</th>
                        <th>使用率(%%)</th>
                        <th>可用容量</th>
                        <th>总容量</th>
                        <th>使用率(%%)</th>
                    </tr>
                </thead>
                <tbody>
                    %s
                </tbody>
            </table>
```

```
            </div>

            <hr />
            <div class="error">请及时处理本该报警信息</div>
        </div>
    </body>

</html>"""

rowTemplate = """
    <tr>
        <td style="border: 1px solid black;border-collapse: collapse;">%s</td>
        <td style="border: 1px solid black;border-collapse: collapse;">%s</td>
        <td style="border: 1px solid black;border-collapse: collapse;">%s</td>
        <td style="border: 1px solid black;border-collapse: collapse;">%s</td>
        <td style="border: 1px solid black;border-collapse: collapse;">%s</td>
        <td style="border: 1px solid black;border-collapse: collapse; color: red; font-weight: bold">%s</td>
        <td style="border: 1px solid black;border-collapse: collapse;">%s</td>
        <td style="border: 1px solid black;border-collapse: collapse; color: red; font-weight: bold">%s</td>
    </tr>
"""


# 发送邮件
def sendMail(host, port, user, password, sender, recipient, mail):

    smtp = smtplib.SMTP()
    smtp.connect(host, port)
    smtp.login(user=user, password=password)
    smtp.sendmail(sender, recipient, mail.as_string())
    smtp.quit()

    return None

# 生成邮件内容
def generatedHTMLMail(mail_content, sender, recipient, subject, cc=None):
```

```python
    msg = htmlTemplate % mail_content

    mail = MIMEMultipart('mixed')
    mail['Accept-Language'] = 'zh-CN'
    mail['Accept-Charset'] = 'ISO-8859-1,utf8'
    mail['From'] = sender
    if isinstance(recipient, list):
        mail['To'] = ';'.join(recipient)
    else:
        mail['To'] = recipient
    mail['Subject'] = subject
    if cc:
        if isinstance(cc, list):
            mail['Cc'] = ','.join(cc)
        else:
            mail['Cc'] = cc

    mail.attach(MIMEText(msg, 'html', 'utf8'))

    return mail

# 生成 HTML 格式的邮件正文
def generatedMailContent(waring_msg):
    content = []
    if isinstance(waring_msg, list):
        for item in waring_msg:
            if item["is_warning"]:
                row = rowTemplate % (item['hostname'], item['ip_address'],
item['mount'], item['inode_available'],
                                    item['inode_total'], item['inode_usage_%'],
item['size_available'],
                                    item['size_total'], item['size_usage_%'])
                content.append(row)

    if len(content) > 0:
        body = '\n'.join(content)
    else:
```

```python
        body = None

    return body

# Ansible module 主函数
def run_module():

    module = AnsibleModule(
        argument_spec=dict(
            smtp_server=dict(required=True, type='str'),
            smtp_port=dict(default=25, type='int'),
            smtp_user=dict(required=True, type='str'),
            smtp_password=dict(required=True, no_log=True, type='str'),
            smtp_sender=dict(required=True, type='str'),
            subject=dict(required=True, type='str'),
            source_data=dict(required=True, type='list')
        ),
        supports_check_mode=True
    )

    result = {
        "changed": False,
        "result": list()
    }

    smtp_server, smtp_port, smtp_user, smtp_password, smtp_sender, mail_subject, source_data = \
        (module.params['smtp_server'], module.params['smtp_port'], module.params['smtp_user'],
         module.params['smtp_password'], module.params['smtp_sender'], module.params['subject'],
         module.params['source_data'])

    send_result = []
    data = {}

    params = dict(
        server=smtp_server,
```

```python
        port=smtp_port,
        user=smtp_user,
        sender=smtp_sender,
        subject=mail_subject,
    )

    try:
        for item in source_data:
            value = dict()
            content = generatedMailContent(item['mounts'])
            if content:
                receiver = item['manager'].strip()
                cc_recipient = None

                tmp = dict(
                  mail_content=content,
                  sender=smtp_sender,
                  recipient=receiver,
                  subject=mail_subject
                )
                mail = generatedHTMLMail(**tmp)

                receivers = list()
                if cc_recipient:
                    if isinstance(receiver, list):
                        receivers.extend(receiver)
                    elif isinstance(receiver, str):
                        receivers.append(receiver)

                    receivers.extend(cc_recipient)

                sendMail(smtp_server, smtp_port, smtp_user, smtp_password, smtp_sender, receivers, mail)

                value.setdefault('recipient', manager)
                value.setdefault('mail', receiver)
                value.setdefault('send', 'ok')
                send_result.append(value)
```

```python
            time.sleep(5000)
        result['changed'] = True
        result.setdefault('result', send_result)
    except Exception:
        result.pop('result')
        result['changed'] = False
        result.setdefault('failed', True)
        result.setdefault('parameters', params)
        result.setdefault('source_data', source_data)
        if data:
            result.setdefault('target_data', data)
        result.setdefault('exception', traceback.format_exc())
    finally:
        module.exit_json(**result)


if __name__ == '__main__':
    run_module()
```

5. Playbook

使用 Playbook 调用相关模块：

```yaml
---
- name: Inspection the linux system mount point information
  hosts: objects
  gather_facts: true
  tasks:
    - check_mountpoints:
        ip_address: "{{ inventory_hostname }}"
        hostname: "{{ ansible_fqdn }}"
        mount_data: "{{ ansible_mounts }}"
        system_alter_value: 90
        application_alter_value: 90
      register: result
    - set_fact:
        cacheable: yes
```

```yaml
      inspection_info: "{{ result.inspection_info }}"

- name: Send mail to mount points's manager
  hosts: sendmail
  gather_facts: false
  tasks:
    - name: Load fact data form caching file
      local_action: read_data
      args:
          cachefile_path: /root/ansible.monitor.mountpointer/caching
      register: result

    - name: send e.mail to manager
      sendmail:
        smtp_server: smtp.163.com
        smtp_port: 25
        smtp_user: test
        smtp_password:
        smtp_sender: test@163.com
        subject: "通知邮件(请勿回复) - Linux 操作系统挂载点使用异常清单"
        source_data: "{{ result.data }}"
```

4.3 操作系统安全基线检查

由于不同用户管理维度的差异，操作系统安全基线检查的项目也有所差异。但总体上的检查项目有十几项，而且因为历史原因用户积累了很多相关的 shell 脚本，下面针对几个常用的检查项目，梳理出较为有效的 Playbook 实现方式。

4.3.1 任务目标

将 Ansible Playbook 和系统中现有的存量 shell 脚本进行结合，灵活实现对 CentOS 的安全基线检查工作。

4.3.2 任务分析

首先，利用现有的存量 shell 脚本实现自动化安全基线检查的目标，必然涉及 shell 脚本的分发和执行结果的回收两个过程。

其次，根据操作系统版本的不同，shell 脚本在检查某些项目时使用的命令会有所差异，所以需要根据 fact 中被管理节点的服务器信息进行版本判断。

最后，为了实现安全基线项目的灵活可配置，需要对 shell 脚本进行拆分。

4.3.3 任务的实现

1. shell 脚本的修改

每个检查项可以根据将检查结果划分为 3 种状态：成功（状态：0）、失败（状态：1）、未检测（状态：2）。例如，确认 SELinux 是否被禁用：

```
#/bin/sh

if ! test -f /etc/selinux/config ; then exit 2 ; fi

if getenforce |grep -i disabled ; then
   exit 0
else
   exit 1
fi
```

2. Playbook——调用 shell 脚本调配模块

使用 Playbook 调用 shell 脚本的示例：

```
- name: Clean UP
  file:
    dest: /tmp/base_line_checkmod_current.sh
    state: absent

- name: Copy mod {{ item_name }} if CentOS 6
  copy:
    src: modules/6/{{ item_file }}
    dest: /tmp/base_line_checkmod_current.sh
    owner: root
    group: root
    mode: 0755
```

```yaml
      changed_when: false
      failed_when: false
      when:
        - ansible_os_family == "CentOS"
        - ansible_distribution_major_version == "6"

    - name: Copy mod {{ item_name }} if CentOS 7
      copy:
        src: modules/7/{{ item_file }}
        dest: /tmp/base_line_checkmod_current.sh
        owner: root
        group: root
        mode: 0755
      changed_when: false
      failed_when: false
      when:
        - ansible_os_family == "CentOS"
        - ansible_distribution_major_version == "7"

    - name: Run Checking Module {{ mod_name }}
      shell: "/tmp/base_line_checkmod_current.sh"
      register: result
      changed_when: false
      failed_when: false
      when:
        - ansible_os_family == "RedHat"

    - name: "check {{ mod_name }} & Report True"
      shell: "echo {{ inventory_hostname }},{{ mod_name }},True >> {{ output }}"
      when: result.rc == 0

    - name: "check {{ mod_name }} & Report False"
      shell: "echo {{ inventory_hostname }},{{ mod_name }},False >> {{ output }}"
      when: result.rc == 1

    - name: "check {{ mod_name }} & Report N/A"
      shell: "echo {{ inventory_hostname }},{{ mod_name }},N/A >> {{ output }}"
      when: result.rc == 2
```

3. Playbook——主任务配置模块

配置主任务模块：

```yaml
---
- hosts: all
  vars:
    output: /tmp/baseline-check-result.txt

  tasks:
  - name: purge output file
    shell: "touch {{ output }}"

  - name: Check SELINUX_IS_DISABLE
    set_fact:
      mod_file: selinux_is_disabled
      mod_name: selinux_is_disabled
  - include : include/module.yml

......

  - name: fetch results {{ inventory_hostname }}
    fetch:
      src: "{{ output }}"
      dest: results/result-{{ inventory_hostname }}.csv
      flat: yes

  - name: clean up
    file:
      dest: "{{ item }}"
      state: absent
    with_items:
      - /tmp/baseline-check-result.txt
      - /tmp/base_line_checkmod_current.sh
```

从上面的 Playbook 中可以看到远端被管理节点上的检测结果文件被回收到控制节点上时默认为文件的后缀名添加了 csv，csv 是一种以纯文本形式存储表格数据（数字和文本）的文件格式，这种格式可以直接被 Excel 等工具打开并展示结果信息。如果需要整合成一个完整的任务报告，则可以自行编写一个简单的 Python 脚本将任务结果汇总成一个 Excel 文件。

因篇幅有限，这里就不再展示完整的安全基线检查代码，还需要读者结合自身的实际需求对本节提到的相关 Playbook 进行完善和丰富。

4.4 收集被管理节点信息

4.4.1 任务目标

收集所有被管理节点的基本信息，通过 template 模块将基本信息渲染成静态 HTML 文件，文件生成后，将文件回收到管理节点的指定路径备用。

4.4.2 任务分析

首先，针对被管理节点的信息，在 Playbook 中设置 gather_facts 为 true 就能自动完成服务器信息的采集，并在任务中使用服务器的基础配置信息。

其次，准备一个 HTML 的模板文件，根据需求使用 Jinja2 将模板中需要填写的信息补充完整。

最后，通过 template 模块将模板文件分发到被管理节点并渲染成最终的 HTML 文件，使用 fetch 模块将渲染后的 HTML 文件收集回控制节点。

4.4.3 Jinja2 简介

Jinja2 是一种现代且设计友好的 Python 模板语言，最初的开发灵感来自 Django 的模板引擎，并在其基础上扩展了语法和一系列强大的功能。最大的亮点是增加了沙箱执行功能和强大的 HTML 自动转义系统，从而提高了 Jinja2 模板文件的安全性。

Jinja2 文件基于 Unicode 编码格式，可以在 Python 2.4 及以上版本中运行（包括 Python 3）。由于 Jinja2 功能特性较多，本节仅介绍其常用的语法规则。

1. 基本语法控制结构

在 Jinja2 中，常见的语法控制结构有以下 3 种。

- 控制结构：{% ... %}；
- 变量取值：{{ ... }}；
- 注释说明：{# ... #}。

2. 变量

模板的变量由传递给模板的上下文字典定义。需要注意的是，"{{}}"不是变量的一部分，在模板文件中相当于 print 语句。

可以使用"['属性名']"的方式访问变量中的属性或元素，也可以使用"."访问变量的属性或元素。示例如下，两种方式的性质完全是一样的：

```
{{ foo.bar }}
{{ foo['bar'] }}
```

如果变量或属性不存在，则返回一个未定义的值。如何处理这种值取决于应用程序配置：
- 默认行为是在输出或迭代时设置变量为空字符串；
- 其他操作都失败。

3. 过滤器

在 Jinja2 中过滤器可以被理解为内置函数和字符串处理函数。变量可以配合过滤器使用以达到修改变量内容的目的。

表 4-2 列举了常用的过滤器。

表 4-2

过滤器名	说明
capitialize	将变量值的首字母转换为大写形式，其他字母转换为小写形式
lower	将变量值转换为纯小写形式
upper	将变量值转换为纯大写形式
title	将变量值中每个单词的首字母都转换为大写形式
trim	将变量值中的首尾空格去掉
join	将多个变量值拼接为新的字符串
replace	替换字符串的值
round	默认对数字进行四舍五入，也可以使用参数进行控制
int	将变量值转换为整型值

过滤器的使用方法如下：

```
{{ 'abc' | captialize }}
# Abc

{{ 'abc' | upper }}
# ABC
```

```
{{ 'hello world' | title }}
# Hello World

{{ 'hello world' | replace('world','test') | upper }}
# HELLO TEST

{{ 18.18 | round }}
# 18.2

{{ 18.18 | round | int }}
# 18
```

想要了解更多 Jinja2 自带的过滤器信息，可以访问其官方网站。

4. 循环控制

在 Jinja2 中可以使用 for 循环对 Python 中的 list、dict 和 tuple 类型进行迭代。使用一对{% for ... %}和{% endfor %}的组合进行语法控制。

迭代 list：

```
<div>
{% for user in users %}
<li>{{ user.username|title }}</li>
{% endfor %}
</div>
```

迭代 dict：

```
<div>
{% for key, value in dict_demo.iteritems() %}
<a>{{ key }}</a>
<a>{{ value }}</a>
{% endfor %}
</div>
```

注意：Jinja2 中不存在 while 循环。

5. 条件语句

Jinja2 中的 if 语句与 Python 中的 if 语句很类似，但需要使用"{% endif %}"表示一个判断语句块的结束。

单一判断：

```
{% if users %}
<ul>
{% for user in users %}
    <li>{{ user.username|e }}</li>
{% endfor %}
</ul>
{% endif %}
```

多分支判断：

```
{% if kenny.A %}
    进入条件 A
{% elif kenny.B %}
    进入条件 B
{% else %}
    未进入条件 A 和条件 B
{% endif %}
```

6. 宏

Jinja2 中的宏可以和常规编程语言的函数功能相提并论，它有助于提升代码的可复用性。定义一个宏需要使用一对"{% macro 宏名(参数)%}"和"{% endmacor %}"的组合来表示宏代码块。具体语法结构示例如下：

```
{% macro input(name, value='', type='text', size=20) -%}
    <input type="{{ type }}" name="{{ name }}" value="{{
        value|e }}" size="{{ size }}">
{%- endmacro %}
```

调用宏时的方式也和常规编程语言类似，调用示例如下：

```
<p>{{ input('username') }}</p>
<p>{{ input('password', type='password') }}</p>
```

在宏内部，可以使用三个特殊变量，如表 4-3 所示。

表 4-3

变量名	说明
varargs	如果传递给宏的位置参数多于宏所能接受的位置参数，则多出的参数会以列表的形式存放在一个名称为 varargs 的变量中
kwargs	类似于 varargs，但针对关键字参数。所有未使用的关键字参数都存储在此特殊变量中
caller	如果宏是从一个调用标记处调用的，那么调用者将作为一个可调用宏存储在这个变量中

在 Jinja2 中，还公开了部分宏相关的属性供模板文件使用，这些属性如表 4-4 所示。

表 4-4

属性名	说明
name	宏的名字。例如 {{ input.name }}，则 input 将被打印出来
arguments	宏收到的参数名称的元组
defaults	宏设置的默认值元组
catch_kwargs	当宏收到额外的关键字参数时返回 true
catch_varargs	当宏收到额外的位置参数时返回 true
caller	如果宏访问特殊的调用者变量并可以从调用标签中调用它，则返回 true

注意：
宏允许跨模板调用，但需要提前将相关的模板导入当前文件。具体导入方法可参考官方文档：【链接 10】。
另外，需要特别注意是，宏名称以 "_" 开头，既不能被导出，也不能被导入。

7. 宏的特殊调用

在某些情况下，将宏传递给另一个宏使用可能会很方便。为此，可以使用特殊调用块。特殊调用块需要使用一对 "{% call %}" 和 "{% endcall %}" 的组合来表示。下面的示例定义了一个宏，并利用该宏展示了特殊调用块的定义及如何使用它：

```
{% macro render_dialog(title, class='dialog') -%}
    <div class="{{ class }}">
        <h2>{{ title }}</h2>
        <div class="contents">
            {{ caller() }}
        </div>
    </div>
{%- endmacro %}
```

```
{% call render_dialog('Hello World') %}
    这是用 render_dialog 函数渲染的内容
{% endcall %}
```

另外，可以将指定参数传递回特殊调用块，使得它可以代替循环逻辑。一般来说，特殊调用块的工作原理与没有名称的宏完全相同（类似于常规编程语言中的匿名函数）。

下面是一个将调用块与参数一起使用的示例：

```
{% macro dump_users(users) -%}
    <ul>
    {%- for user in users %}
        <li><p>{{ user.username|e }}</p>{{ caller(user) }}</li>
    {%- endfor %}
    </ul>
{%- endmacro %}

{% call(user) dump_users(list_of_user) %}
    <dl>
        <dl>Realname</dl>
        <dd>{{ user.realname|e }}</dd>
        <dl>Description</dl>
        <dd>{{ user.description }}</dd>
    </dl>
{% endcall %}
```

8. 模板和模板继承

Jinja2 最强大的功能是模板的继承。用户可以先构建一个基础模板，该模板包含站点的所有常见元素，并定义子模板可以覆盖的块。是不是很像常规面向对象开发语言中关于对象的继承和使用？下面用详细示例说明模板继承的实际应用情况。

1）基础模板

假设我们创建了一个名为 base.html.j2 的模板文件并定义了一个简单的 HTML 文本框架模板。模板文件中定义了一个简单的 HTML 页面，分为 head 和 body 两个空白部分，子模板的作用就是填充这两部分的内容：

```
<!DOCTYPE html>
<html lang="en">
```

```
<head>
    {% block head %}
    <link rel="stylesheet" href="style.css" />
    <title>{% block title %}{% endblock %} - My Webpage</title>
    {% endblock %}
</head>
<body>
    <div id="content">{% block content %}{% endblock %}</div>
    <div id="footer">
        {% block footer %}
        &copy; Copyright 2021 by <a href="http://www.sinontt.com/">sinontt</a>.
        {% endblock %}
    </div>
</body>
</html>
```

可以看到"{% block %}"和"{% endblock %}"一一对应构成一个可填充的块，整个基础模板包含四个可填充部分。

注意：
block 标签可以被应用到其他代码块中（如 if 代码块），但 block 标签部分的内容总是被执行，而不管 if 代码块是否需要实际呈现出来。

2）子模板

```
{% extends "base.html.j2" %}
{% block title %}hello world{% endblock %}
{% block head %}
    {{ super() }}
    <style type="text/css">
        .important { color: #336699; }
    </style>
{% endblock %}
{% block content %}
    <h1>Title 1</h1>
    <p class="important">
      Welcome to my awesome homepage.
```

```
    </p>
{% endblock %}
```

这里能看到一个新标签"{% extends %}",它是整个子模板的关键部分,它告诉模板引擎这个模板扩展了另一个模板。当模板系统渲染此模板时,它会尝试定位父模板。需要注意的是,extends 标签必须是子模板中的第一个标签。

基础模板的文件名取决于模板加载器,例如,FileSystemLoader 允许通过提供文件名访问其他模板。Jinja2 允许使用带有"/"的子目录方式访问基础模板,例如:

```
{% extends "templates/default.html.j2" %}
```

另外,需要注意的是,上面的子模板中没有定义 footer 对应的块,所以将使用父模板中的值。

注意:
在同一个模板中不能定义多个具有相同名称的"{% block %}"标签。存在这一限制的原因是,块标签在"两个方向"上均起作用。也就是说,块标签不仅提供了要填充的占位符,还定义了在父元素中填充占位符的内容。如果一个模板中有两个命名为"{% block %}"的标签,则该模板的父模板不知道使用哪个代码块的内容。
如果需要多次打印一个代码块,则可以使用特殊的 self 变量并使用该名称调用块,例如:

```
<title>{% block title %}{% endblock %}</title>
<h1>{{ self.title() }}</h1>
{% block body %}{% endblock %}
```

9. HTML 转义

在正常的开发环境中,使用模板文件生成 HTML 文件时,始终存在由于传入的变量信息最终影响 HTML 显示风格的问题。为了解决这样的问题,一般有两种解决方案:

- 手动转义每个变量;
- 默认情况下自动转义所有内容。

Jinja2 同时支持这两种解决方案,使用哪种解决方案取决于应用程序的配置。默认配置为不自动转义。这是由于以下两个原因导致的:

- 转义除安全值外的所有内容,这也意味着 Jinja2 转义了所有已知不包含 HTML 元素的变量(如数字、布尔值),这可能导致严重的性能问题;
- 有关变量安全性的信息非常脆弱。通过强制使用安全值和非安全值,返回值可能是被经过二次转义后的 HTML 文件。

1）手动转义

如果启用了手动转义，则需要自行判断何时对变量进行转义，以及要转义什么？如果一个变量中包含以下任何四种字符：">""<""&"或"""，那么除非保证该变量包含结构良好的、值得信赖的 HTML 结构，否则应该手动转义该变量。通过将变量和"|e"过滤器结合来传递数据的方式实现手动转义，例如：

{{ user.username|e }}

2）自动转义

启用自动转义后，默认情况下所有内容都将自动转义，除非是明确标记为安全的值。变量和表达式可以在以下两种情况中被标记为"安全"：

- 应用程序提供的上下文字典"markupsafe.Markup"；
- 带有"|safe"过滤器的模板。

如果标记为安全的字符串通过其他无法理解该标记的 Python 代码传递，则可能导致标记丢失。在到达模板之前，请注意何时将数据标记为安全数据，以及如何对其进行处理。

如果一个值已被转义但未标记为"安全"，则仍将自动转义并生成二次转义的字符。如果已经明确知道存在安全数据但未对数据进行标记，则确保将其包装在 Markup 标签中或使用"|safe"过滤器对数据进行安全处理。

Jinja2 的函数功能模块（macros、super、self.BLOCKNAME）始终返回标记为安全的模板数据。而带有自动转义的模板中的字符串变量被认为是不安全的，因为原生 Python 字符串（str、unicode、basestring）不是安全的，需要用 __html__ 属性标记字符串。

4.4.4 服务器巡检任务

对服务器进行巡检是运维工程师的日常工作之一，我们可以编写巡检任务的文件，让 Ansible 为我们完成自动化的服务器巡检任务。

1. 目录结构

巡检任务的 Playbook 的目录结构如下：

```
├── Inventory
│   └── hosts
├── Pipfile
├── Pipfile.lock
```

```
├── ansible.cfg
├── ansible.module.demo.py
├── baseinformation.yml
├── library
│   ├── grub2info.py
│   ├── manufacture.py
│   ├── memory.py
│   ├── network.py
│   ├── runlevel.py
│   ├── uptime.py
│   └── yumcheckupdate.py
└── templates
    ├── system_health_check.css
    ├── system_health_check.html.j2
└── system_health_check.js
```

2. Playbook

关键巡检任务的配置如下：

```
---
- name: Get basic information about the operating system
  hosts: all
  gather_facts: true
  tasks:
    - name: Get linux system run level
      runlevel:
      register: system_runlevel

    - name: Get linux grub menus information
      grub2info:
      register: grub_menus

    - name: Get system start time
      shell: date -d "$(awk -F. '{print $1}' /proc/uptime) second ago" +"%Y-%m-%d %H:%M:%S"
      register: starttime

    - name: Get system uptime
      uptime:
```

```yaml
    register: sys_uptime

  - name: Get dmi information about BIOS, Baseboard, System
    manufacture:
    register: dmidecode

  - name: Get memory information
    memory:
    register: sys_memory

  - name: Get graphics card information
    graphics:
    register: sys_graphics

  - name: Get timestap
    shell: date +%Y%m%d%H%M%S.%N
    register: timestap

  - name: Padding data to HTML file
    template:
      src: system_health_check.html.j2
      dest: /tmp/{{inventory_hostname}}-{{ansible_fqdn}}-{{timestap.stdout}}-system_health_check.html
    when: timestap.rc == 0
```

4.5 小结

在实际工作中有大量的自动化运维需求，编写 Playbook 需要适当地理解如何使用变量，变量如何存在于内存或 fact 中。结合 Ansible 提供的标准 module 可以完成大部分自动化运维任务，当 Ansible 的标准 module 不能满足任务需求时，可以使用自定义 module、自定义 fact module 或自定义 info module 来实现所需的功能。需要额外注意的是，理解 Ansbile module 和 plugin 之间的区别和针对的具体对象，才能正确地使用 Ansible 自定义功能更好地为任务目标服务。最后，读者可能还需要对 Python 有一定的了解才能更好地使用 Ansible Playbook。

第 5 章 AIOps 概述

5.1 AIOps 概述

2013 年，Gartner 在一份报告中提及了 ITOA，当时的定义是 IT 运营分析（IT Operations Analytics），通过技术与服务手段，采集、存储、展现海量的 IT 运维数据，并进行有效的推理与归纳得出分析结论。

Gartner 在 2015 年对 ITOA 应该具备的能力进行了定义：

（1）ML/SPDR：机器学习/统计模式发现与识别。
（2）UTISI：非结构化文本索引，搜索以及推断。
（3）Topological Analysis：拓扑分析。
（4）Multi-dimensional Database Search and Analysis：多维数据库搜索与分析。
（5）Complex Operations Event Processing：复杂运维事件处理。

而随着时间推移，在 2016 年，Gartner 将 ITOA 概念升级为了 AIOps，原本的含义基于算法的 IT 运维（Algorithmic IT Operations），即平台利用大数据、现代的机器学习技术和其他高级分析技术，通过主动、个性化和动态的洞察力直接或间接地、持续地增强 IT 操作（监控、自动化和服务台）功能。AIOps 平台可以同时使用多个数据源、多种数据收集方法、实时分析技术、深层分析技术及展示技术。

Gartner 对 AIOps 的能力的定义：

（1）Historical data management：历史数据管理。
（2）Streaming data management：流数据管理。
（3）Log data ingestion：日志数据整合。
（4）Wire data ingestion：网络数据整合。
（5）Metric data ingestion：指标数据整合。
（6）Document text ingestion：文本数据整合。
（7）Automated pattern discovery and prediction：自动化模式发现和预测。
（8）Anomaly detection：异常检测。
（9）Root cause determination：根因分析。
（10）On-premises delivery：提供私有化部署。
（11）Software as a service：提供 SaaS 服务。

随着 AI 在多个领域越来越火爆，Gartner 也在 2017 年年中一份报告中，将 AIOps 的含义定义为 Artificial Intelligence for IT Operations，也就是智能化运维。

5.2 AIOps 的落地路线

虽然 AIOps 的概念被提出来有一段时间了，并且 Gartner 也预测到 2022 年，约 40%的大型企业将部署 AIOps 平台，也提出了 AIOps 应该具备的能力，但是到目前为止，AIOps 并不像 DevOps 一样，已经具备一个非常完整的最佳实践的形态，以至于每个人心目中的 AIOps 可能都是不一样的。可以这样说，AIOps 在算法层面已经有一定的方向，但是在工程化层面上，还处于百花齐放的状态。

由于还没有一个具体可参考的对象，人们自然而然地充分发挥了各自的创造力，出现了多种风格的 AIOps 工程化落地路线。总体来看，目前主要有以下几种工程化落地路线。

- 时序指标路线：选择这个路线的团队前身大多是以监控系统为主的，由于在监控系统中积累了大量的时序数据，结合监控系统的日常工作特性，他们选择将单指标时序预测、多指标时序预测、单指标异常检测、多指标异常检测、指标关联分析等以时序指标分析为主的方法整合进系统中。这条落地路线主要解决的问题是让监控系统能够自适应地告警，通过异常检测触发运维动作，通过指标关联分析辅助决策来缩短解决故障定位的时间。
- 事件分析路线：选择这个路线的团队以非结构化数据分析为主，以日志分析系统、告警中心的团队多见。所使用的关键技术包括告警事件降噪、事件发现、告警事件抑制、日志聚类、事件解决方案推荐等。相比时序指标路线，事件分析路线的团队认为一切的运维数据最终都会反映成运维事件，以运维事件的角度去分析才是 AIOps 的最终落地解决方案。

- 知识增强路线：这个路线的团队认为运维中起决定作用的是知识，通过整理运维知识，形成运维知识图谱，基于知识图谱和知识库来提升运维知识的利用率才是 AIOps 的最终目标，通过使用知识检索、知识推理、命名实体识别等技术，结合运维知识库、基础数据，形成运维机器人，在原有的知识库基础上赋予 AI 的能力，最终达到 AIOps 的目标。
- AI 平台路线：这个路线的团队认为 AI 能力不应该直接强绑定在运维工具上，应该将 AI 能力沉淀在 AI 平台上，通过 AI 平台赋能已有应用的方式，最终达到 AIOps 的目标。它的特点是让运维系统和 AI 系统在平台层面上进行隔离，通过 API 的方式互相整合。

5.3 基于基础指标监控系统的 AIOps

当团队已经在时序指标上有非常多的积累的时候，可以尝试使用时序指标路线的 AIOps 方案，如图 5-1 所示。

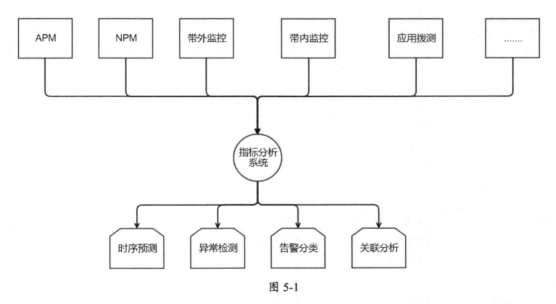

图 5-1

首先采用 APM、NPM、带内监控、带外监控等技术手段尽可能全面地对指标进行收集，在收集了多种类型的指标后，面临的第一个问题是如何高效地对指标进行存储和使用。这里有两种常见的数据存储方案，第一种是采用 Elasticsearch，第二种是采用 OpenTSDB。采用 Elasticsearch 的优势在于 Elasticsearch 的 Schema On Write 的能力，能够让我们灵活地定义数据的存储模式。同时，Elasticsearch 具备大数据的统计查询能力，综合能力表现优秀。而第二种常见存储手段是采用时间序列数据库，比较常见的方式是 OpenTSDB。OpenTSDB 背靠 Hadoop 大数据生态，使得它具备了大数据的存储分析能力，同时精心设计过的时间序列存储方式也提高了其数据分析能力。但是比起 Elasticsearch 和 OpenTSDB，笔者更加推荐使用 ClickHouse 进行时间序列的存储分析。最重要的一点是 ClickHouse 的性能（评测数据来源于【链接 11】），如表 5-1 所示。

表 5-1

场景 1	场景 2	场景 3	场景 4	数据库
0.005	0.011	0.103	0.188	BrytlytDB 2.1 & 5-node IBM Minsky cluster
0.009	0.027	0.287	0.428	BrytlytDB 2.0 & 2-node p2.16xlarge cluster
0.021	0.053	0.165	0.51	OmniSci & 8 Nvidia Pascal Titan Xs
0.027	0.083	0.163	0.891	OmniSci & 8 Nvidia Tesla K80s
0.028	0.2	0.237	0.578	OmniSci & 4-node g2.8xlarge cluster
0.034	0.061	0.178	0.498	OmniSci & 2-node p2.8xlarge cluster
0.036	0.131	0.439	0.964	OmniSci & 4 Nvidia Titan Xs
0.051	0.146	0.047	0.794	kdb+/q & 4 Intel Xeon Phi 7210 CPUs
0.134	0.349	0.542	3.312	OmniSci & a 16" MacBook Pro
0.241	0.826	1.209	1.781	ClickHouse, 3 x c5d.9xlarge cluster
0.466	1.094	0.742	1.412	Hydrolix & c5n.9xlarge cluster
0.762	2.472	4.131	6.041	BrytlytDB 1.0 & 2-node p2.16xlarge cluster
1.034	3.058	5.354	12.748	ClickHouse, Intel Core i5 4670K
1.56	1.25	2.25	2.97	Redshift, 6-node ds2.8xlarge cluster
2	2	1	3	BigQuery
2.362	3.559	4.019	20.412	Spark 2.4 & 21 x m3.xlarge HDFS cluster
3.54	6.29	7.66	11.92	Presto 0.214 & 21 x m3.xlarge HDFS cluster
4	4	10	21	Presto, 50-node n1-standard-4 cluster
4.88	11	12	15	Presto 0.188 & 21-node m3.xlarge cluster
6.41	6.19	6.09	6.63	Amazon Athena
8.1	18.18	n/a	n/a	Elasticsearch (heavily tuned)
10.19	8.134	19.624	85.942	Spark 2.1, 11 x m3.xlarge HDFS cluster
11	10	21	31	Presto, 10-node n1-standard-4 cluster
11	14	16	22	Presto 0.188 & single i3.8xlarge w/ HDFS
14.389	32.148	33.448	67.312	Vertica, Intel Core i5 4670K
22	25	27	65	Spark 2.3.0 & single i3.8xlarge w/ HDFS
28	31	33	80	Spark 2.2.1 & 21-node m3.xlarge cluster
34.48	63.3	n/a	n/a	Elasticsearch (lightly tuned)
35	39	64	81	Presto, 5-node m3.xlarge HDFS cluster

续表

场景 1	场景 2	场景 3	场景 4	数据库
43	45	27	44	Presto, 50-node m3.xlarge cluster w/ S3
152	175	235	368	PostgreSQL 9.5 & cstore_fdw
264	313	620	961	Spark 1.6, 5-node m3.xlarge cluster w/ S3
448	797	1811	3286	SQLite 3, Parquet & HDFS
1103	1198	2278	6446	Spark 2.2, 3-node Raspberry Pi cluster
31193	NR	NR	NR	SQLite 3, Internal File Format

场景 1:

```
SELECT cab_type,
       count(*)
FROM trips
GROUP BY cab_type;
```

场景 2:

```
SELECT passenger_count,
       avg(total_amount)
FROM trips
GROUP BY passenger_count;
```

场景 3:

```
SELECT passenger_count,
       extract(year from pickup_datetime) AS pickup_year,
       count(*)
FROM trips
GROUP BY passenger_count,
         pickup_year;
```

场景 4:

```
SELECT passenger_count,
       extract(year from pickup_datetime) AS pickup_year,
       cast(trip_distance as int) AS distance,
       count(*) AS the_count
FROM trips
```

```
GROUP BY passenger_count,
         pickup_year,
         distance
ORDER BY pickup_year,
         the_count desc;
```

可以看到，在 OLAP 领域，作为开源数据库的 ClickHouse 的性能非常高，拥有高性能的 OLAP 数据库对实现 AIOps 非常有帮助，能够极大地缩短 AIOps 实现过程中的数据分析与模型训练的时间。完成数据采集与存储后，对原有的监控系统流程进行升级改造。基于基础监控的 AIOps 实践如图 5-2 所示。

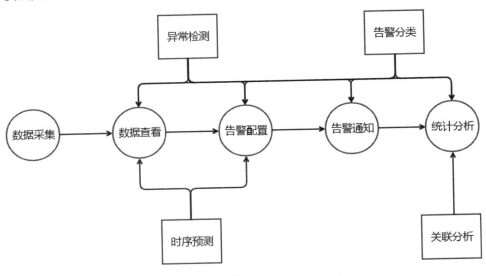

图 5-2

基于基础指标监控系统的 AIOps 落地方案的常见步骤如下：

（1）将时序预测功能接入数据查看和告警配置功能模块，能够为监控系统加入动态告警阈值的能力，同时还能在可视化界面提供指标发展趋势的曲线。

（2）将异常检测功能接入数据查看、告警配置、统计分析功能模块，让系统具备异常数据发现的能力，提升监控系统的监控能力。

（3）将告警分类功能接入告警通知功能模块，提升监控系统的告警分组能力，能够对告警进行分组合并，提升告警处理的效率。

（4）将关联分析功能接入统计分析模块，让系统获得关联分析的能力，能够给出哪些指标的变化是与待分析指标有关联的，帮助运维工程师减少需要分析指标的数量。

5.4 基于日志分析系统的 AIOps

当团队已经具备比较成熟的日志分析平台时，可以尝试在日志分析平台中加入 AIOps 的能力。由于非结构化数据所带来的信息含量更多，所以日志分析平台可改动的点也会更多。基于日志分析系统的 AIOps 方案概览如图 5-3 所示。

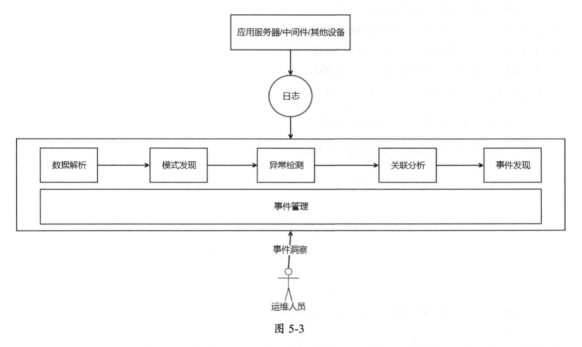

图 5-3

首先，我们可以采用日志采集 Agent 对应用、服务器、中间件的日志进行采集，采集完后将日志输入日志分析平台，日志分析平台会对日志进行索引，之后就可以对日志进行检索和分析了，这是日志分析平台的基本功能，接下来进行平台升级。

（1）引入事件管理模块：将事件的告警、CEP（复杂事件处理）所得到的信息全部转储到事件索引中，并且对事件进行结构化处理，提取出对象名称、创建时间、事件信息等关键字段。读者可能会感到疑惑，为什么要将日志转换为事件呢？这一切都是为了将多维度的数据转换到同一个事件维度，并且对数据量进行压缩。举个例子，最近 5 分钟出现了 100 条 DataBase Connection Refuse 日志，通过将日志转换为事件，可以得到一条数据库连接失败的事件。

（2）在日志数据格式化阶段，引入日志分类的能力，在可视化配置日志解析规则的时候，能够提示此类日志使用哪种正则表达式或者 Grok 表达式进行格式化，降低数据格式化规则的配置难度。

（3）在日志检索模块中引入日志泛化功能，通过日志聚类的功能实现日志智能聚类的能力，即将相似的日志进行合并，把变化的地方用星号替换，把变动的地方进行合并，这样能够极大地降低日志分析时的阅读成本。

假设日志原文如下：

Temperature (41C) exceeds warning threshold
Temperature (41C) exceeds warning threshold
Temperature (41C) exceeds warning threshold
Temperature (41C) exceeds warning threshold

经过日志泛化功能处理后，得到如下结果：

Temperature (* C) exceeds warning threshold 4

基于日志分析的 AIOps 实践如图 5-4 所示。

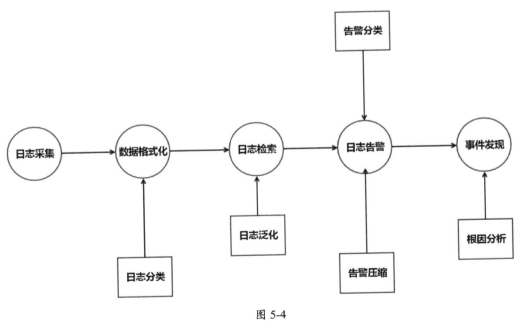

图 5-4

（1）在日志告警流程中引入告警分类和告警压缩的能力，提升平台的告警抑制能力。

（2）在事件发现模块引入告警根因分析模块，通过告警根因分析算法找出告警的根因，以及告警的传播链，提升运维工程师分析告警根因的效率。

5.5 基于知识库的 AIOps

随着业务的发展，设备、网络等基础设置的运维知识也在飞速增加，日常运维中会积累非常多的运维知识。运维知识种类如图 5-5 所示。

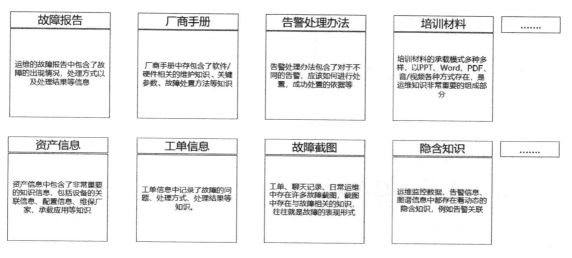

图 5-5

对于以知识管理为核心的团队，可以尝试采用充分利于知识分析的路线进行 AIOps 的落地，在这条路线上，一切的指标、日志最终都反映为知识。如何让知识得到更加充分的利用是关键。在知识库的路线下，我们可以加入文本摘要、文本分类、文本聚类、图谱抽取、语义检索、知识识别、实体识别、实体关系提取的能力来实现升级知识库的目标。基于知识库的 AIOps 实践如图 5-6 所示。

（1）OCR 和语音转写功能作为知识采集的基础 AI 能力，可以使数据源的采集更加广泛。

（2）在知识加工模块中加入文本分类、文本聚类和语义编码的能力，主要用于为后续的知识应用和知识理解做准备，提供训练用的基础素材。

（3）在知识理解模块中加入实体识别和实体关系提取的能力，为后续运维机器人执行运维操作做准备。

（4）经历了前面的一系列应用，提升运维效率的功能都可以加载在知识应用的模块中，一是提供基于语义的检索能力，让运维知识更加容易被检索出来；二是提供运维机器人的能力，知识库能够通过自然语言理解操作者的意图；三是提供实体识别、实体关系提取、文本相似度检索能力，联动 Ansible 完成自动化运维的操作。

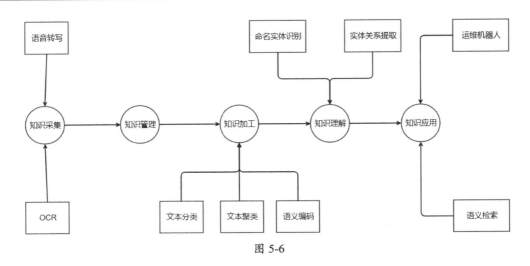

图 5-6

5.6 基于 AI 平台的 AIOps

AI 平台落地路线是笔者比较推荐的一条落地路线，最主要的原因是 AI 平台将 AI 能力进行统一管理，与具体的应用系统分离。这非常好地解决了 AIOps 没有一个最佳实践的问题，通过模块化 AI 能力接入的方式，可以极大地提升 AI 能力的复用性，现有的运维系统不再需要实现 AI 模型生产所必需的功能（包括数据管理、数据标注、模型管理、模型训练、显卡资源调度、AutoML、模型上线等）。基于 AI 平台的 AIOps 实践如图 5-7 所示。

图 5-7

有了 AI 平台后，我们只需要通过 API 的方式，将 AI 能力暴露给应用即可。在这种模式下，应用不再需要关注算法是如何从原始的数据最终变成可使用的模型的，只需要根据自身的需要，从 AI 平台上调取平台所需要的 AI 能力即可，对原有运维系统的改动最小，能力可复用性最大，但是首次投入的成本也会较大。基于 AI 平台的 AIOps 联动方式如图 5-8 所示。

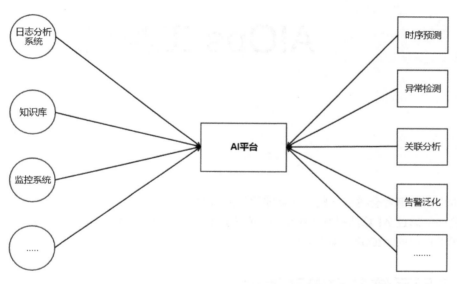

图 5-8

第 6 章
AIOps 工具包

在本章中,笔者会挑选一些"开箱即用"的 AIOps 工具包,让读者能够采用尽量少的代码就能体验如何通过 AI 算法提升运维效率,所选择的工具包以易用和有效为主,期望帮助读者快速构建属于自己的 AIOps 工具包。

6.1 应用系统参数自动优化

应用系统、操作系统都有非常多可调整的参数,选择最佳的参数能够帮助应用系统提升非常多的性能,使得应用系统使用更少的资源表现出更高的性能。但是,这么多参数,该如何调整才是最合适的呢?在没有使用 AIOps 工具之前,通常这个任务会交给有经验的运维工程师,他们基于多年的经验,会在众多参数中选择最能够提升应用系统性能的一小批参数进行调整,在测试环境中进行验证通过后,更新到生产环境中。经过经验丰富的运维工程师进行参数的调整后,应用系统的性能至少能提升 30%~50%。

有没有办法让应用系统自动调整参数呢?假如应用系统能自动调整参数,那么能不能将参数优化做到极致,对这个应用系统的全量参数去做优化,而不只是选择一小部分参数进行优化呢?AIOps 工具包中的参数自动优化工具包正是解决这个问题的工具包。微软开源的深度学习平台 OpenPAI 中有一个参数自动优化组件 NNI,在深度学习中,可以使用 NNI 作为模型的超参数检索工具,自动化地检索 AI 算法中的最佳超参数。这个工具同样可以迁移到运维领域,用自动化参数优化工具来优化应用系统或操作系统。接下来以优化 Kafka 为例介绍 NNI 的使用方法。

Kafka 是一个高性能的消息中间件,假如人工对 Kafka 参数调优,则需要花费不少的时间。

下面用 NNI 对 Kafka 进行参数的自动化优化，找到 Kafka 的最佳参数组合，安装 Kafka 的过程不再赘述，直接进入 NNI 工具包使用的主题。

下面是此次使用的 Kafka 配置文件：

```
broker.id=0
num.network.threads=3
num.io.threads=8
socket.send.buffer.bytes=102400
socket.receive.buffer.bytes=102400
socket.request.max.bytes=104857600
log.dirs=/tmp/kafka-logs
num.partitions=1
num.recovery.threads.per.data.dir=1
offsets.topic.replication.factor=1
transaction.state.log.replication.factor=1
transaction.state.log.min.isr=1
log.retention.hours=168
log.segment.bytes=1073741824
log.retention.check.interval.ms=300000
zookeeper.connect=localhost:2181
zookeeper.connection.timeout.ms=18000
group.initial.rebalance.delay.ms=0
```

其中进行超参数检索的参数为：

- num.network.threads；
- num.io.threads；
- socket.send.buffer.bytes；
- socket.receive.buffer.bytes；
- socket.request.max.bytes；
- num.partitions。

准备好 Kafka 的安装文件后先启动 Kafka，做一些测试前的实验，Kafka 启动完成后，先创建一个 topic：

```
./bin/kafka-topics.sh --bootstrap-server 127.0.0.1:9092 --create --topic test
```

进行一次简单的性能测试：

```
./bin/kafka-producer-perf-test.sh --topic test --num-records 5000 --record-size 100
--throughput -1 --producer-props acks=0 bootstrap.servers=127.0.0.1:9092
```

得到如下输出：

```
5000 records sent, 10638.297872 records/sec (1.01 MB/sec), 2.49 ms avg latency, 383.00
ms max latency, 1 ms 50th, 12 ms 95th, 13 ms 99th, 13 ms 99.9th.
```

我们期望获得的性能计数是 10638.297872，以这个值作为参数检索的性能指标。

一切都准备好后安装 NNI：

```
pip install nni
```

本次性能基准测试固定每次发送 500000 条消息，每条消息大小为 100 字节，关闭生产者的节流操作，期望找到此场景下吞吐量最大的参数组合。

编写 search_space.json 文件，此文件是 NNI 用于检索参数的搜索空间：

```
{
    "num_network_threads": {
        "_type": "randint",
        "_value": [1, 40]
    },
    "num_io_threads": {
        "_type": "randint",
        "_value": [1, 40]
    },
    "socket_send_buffer_bytes": {
        "_type": "randint",
        "_value": [10240, 1024000]
    },
    "socket_receive_buffer_bytes": {
        "_type": "randint",
        "_value": [10240, 1024000]
    },
    "socket_request_max_bytes": {
        "_type": "randint",
        "_value": [10485760, 1048576000]
```

```
        },
        "num_partitions": {
            "_type": "randint",
            "_value": [1, 40]
        }
    }
```

编写测试代码的主函数：

```
import nni
from jinja2 import Template
import time
import os
from logzero import logger


def run(**parameters):
    num_network_threads = parameters['num_network_threads']
    num_io_threads = parameters['num_io_threads']
    socket_send_buffer_bytes = parameters['socket_send_buffer_bytes']
    socket_receive_buffer_bytes = parameters['socket_receive_buffer_bytes']
    socket_request_max_bytes = parameters['socket_request_max_bytes']
    num_partitions = parameters['num_partitions']

    with open('./conf.jinja2', 'r') as file_:
        template = Template(file_.read())
        rendered = template.render(
            num_network_threads=num_network_threads,
            num_io_threads=num_io_threads,
            socket_send_buffer_bytes=socket_send_buffer_bytes,
            socket_receive_buffer_bytes=socket_receive_buffer_bytes,
            socket_request_max_bytes=socket_request_max_bytes,
            num_partitions=num_partitions
        )
    with open('/home/softs/kafka_2.12-2.7.0/config/server.properties', 'w') as f:
        f.write(rendered)
    os.popen('/home/softs/kafka_2.12-2.7.0/bin/kafka-server-start.sh -daemon /home/softs/kafka_2.12-2.7.0/config/server.properties')
```

```python
    logger.info('Kafka 启动成功')
    time.sleep(3)

    logger.info('执行性能测试')
    p = os.popen('/home/softs/kafka_2.12-2.7.0/bin/kafka-producer-perf-test.sh --topic test --num-records 5000 --record-size 100 --throughput -1 --producer-props acks=0 bootstrap.servers=127.0.0.1:9092')
    content = p.read()
    score=content.split(',')[1].strip().split(' ')[0]
    os.popen('/home/softs/kafka_2.12-2.7.0/bin/kafka-server-stop.sh')
    logger.info('Kafka 停止成功')
    return float(score)


def generate_params(received_params):
    params = {
        "num_network_threads": 1,
        "num_io_threads": 8,
        "socket_send_buffer_bytes": 102400,
        "socket_receive_buffer_bytes": 102400,
        "socket_request_max_bytes": 104857600,
        "num_partitions": 1,
    }

    for k, v in received_params.items():
        params[k] = int(v)

    return params


received_params = nni.get_next_parameter()
params = generate_params(received_params)

throughput = run(**params)
nni.report_final_result(throughput)
```

代码的逻辑如下：

（1）从 NNI 中获取本次输入的参数。

（2）将参数转换成 Kafka 的配置文件。

（3）启动 Kafka。

（4）执行 Kafka 生产者性能测试。

（5）获取生产者的每秒消息生产速率。

（6）停止 Kafka。

以上 6 个步骤形成了一个完整的评测用例，剩下的检索工作就可以交给 NNI 去完成了。编写 config_tpe.yml：

```yaml
authorName: default
experimentName: auto_kafka_TPE
trialConcurrency: 1
maxExecDuration: 12h
maxTrialNum: 15
trainingServicePlatform: local
searchSpacePath: search_space.json
useAnnotation: false
tuner:
  builtinTunerName: TPE
  classArgs:
    optimize_mode: maximize
trial:
  command: python3 main.py
  codeDir: .
  gpuNum: 0
```

执行 nnictl create --config ./config_tpe.yml 可以看到如图 6-1 所示的信息，表示启动 NNI 并测试成功。

此时可以查看 NNI 的概览界面，概览界面展示了此次评估执行的次数、执行的时间、最好的评估结果及评估 Top10 的评估结果，NNI 参数优化首页如图 6-2 所示。

单击"Experiment"进入详情界面，Default metric 界面展示了每一次实验的评估结果，可以直观地看到每一次性能评估的变化，如图 6-3 所示。

```
INFO:  expand searchSpacePath: search_space.json to /home/workspaces/kafka-autoturn/search_space.json
INFO:  expand codeDir: . to /home/workspaces/kafka-autoturn/.
INFO:  Starting restful server...
INFO:  Successfully started Restful server!
INFO:  Setting local config...
INFO:  Successfully set local config!
INFO:  Starting experiment...
INFO:  Successfully started experiment!
------------------------------------------------------------------------
The experiment id is KLnOHhko
The Web UI urls are: http://127.0.0.1:8080    http://192.168.199.249:8080    http://172.17.0.1:8080
------------------------------------------------------------------------

You can use these commands to get more information about the experiment
------------------------------------------------------------------------
         commands                      description
    1. nnictl experiment show       show the information of experiments
    2. nnictl trial ls              list all of trial jobs
    3. nnictl top                   monitor the status of running experiments
    4. nnictl log stderr            show stderr log content
    5. nnictl log stdout            show stdout log content
    6. nnictl stop                  stop an experiment
```

图 6-1

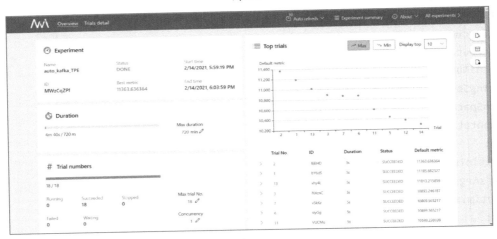

图 6-2

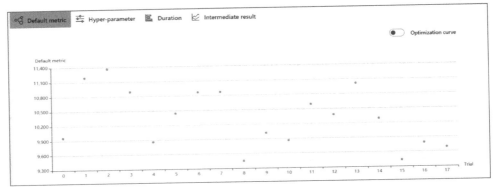

图 6-3

Hyper-parameter 界面展示了每个评估结果所对应的参数，如图 6-4 所示。

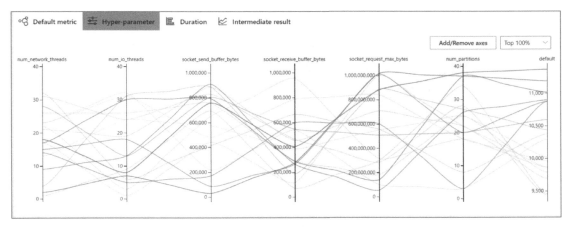

图 6-4

在 Duration 界面中可以查看每次评估执行的时长，如图 6-5 所示。

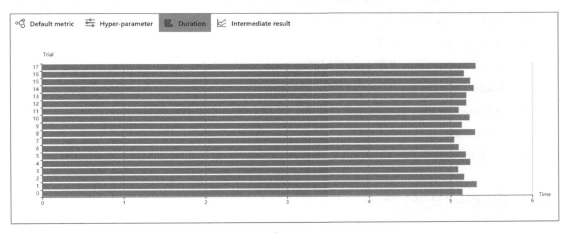

图 6-5

我们可以在 Trial Jobs 界面中查看本次实验检索出来的最佳参数，如图 6-6 所示。

简单介绍一下 search_space.json 和 config_tpe.yml，首先是 search_space.json，这个文件是用来定义每个超参数的检索范围的，用于控制每次评估传入哪些参数，具体的配置可以参考【链接 12】。

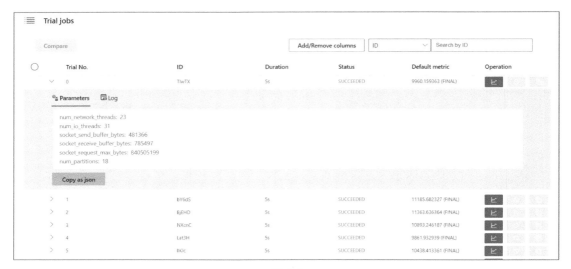

图 6-6

再看另外一份文件 config_tpe.yml，这是启动参数检索的描述文件，其中最重要的参数如下：

- maxExecDuration：最大执行时间；
- maxTrialNum：最大执行次数；
- searchSpacePath：参数检索空间的路径；
- builtinTunerName：优化器的算法名称；
- optimize_mode：优化器的优化方向，是越大越好还是越小越好；
- command：执行评测的脚本。

可选的优化器可以参考【链接 13】。

6.2 智能日志分析

在日常运维过程中，我们会收集非常多的日志信息，日志信息中有非常多有价值的信息，但是期望通过人工去分析全量的日志数据是不现实的。这时就可以使用模式发现算法找出海量日志中的模式。对日志使用模式发现算法，最直接的帮助就是能够通过观察少量的模式分析全量的数据，大大提升了日志分析的效率。更进一步，可以做日志模式的统计分析，发现那些以前没有出现过的日志模式，提醒运维工程师引起关注，也可以对日志模式进行同比和环比分析，观察日志在一个周期内的变化幅度是否正常。

Spell 算法是李飞飞等人在论文 *Spell Streaming Parsing of System Event Logs* 中提出来的，该论文提出了一种在线实时动态解析日志的方法，基于最长公共子序列的方式，提供了实时处理日志数据输入、不断生成新的日志模式的能力。

一般来说，日志都是以一定模式存在的，主要原因是日志是由代码产生的，开发人员在编写代码的时候，都会使用一定的模板。以 Python 为例，在输出日志的时候会编写如下代码，导致日志在产生后本身就是模板化的：

```
logger.info(f'当前用户为:[{user_name}],执行操作:[{action}]')
```

由于日志分析平台存放了海量的日志，通过一个算法自适应地提取日志中的模板是最合适的。换个角度，直接从日志原文来看，也可以非常明显地看出日志是有一定模式的。下面以 Apache 日志为例进行说明：

```
[Sun Dec 04 04:47:44 2005] [notice] workerEnv.init() ok
/etc/httpd/conf/workers2.properties
[Sun Dec 04 04:47:44 2005] [error] mod_jk child workerEnv in error state 6
[Sun Dec 04 04:51:08 2005] [notice] jk2_init() Found child 6725 in scoreboard slot 10
[Sun Dec 04 04:51:09 2005] [notice] jk2_init() Found child 6726 in scoreboard slot 8
[Sun Dec 04 04:51:09 2005] [notice] jk2_init() Found child 6728 in scoreboard slot 6
```

可以看到如下日志：

```
[Sun Dec 04 04:51:08 2005] [notice] jk2_init() Found child 6725 in scoreboard slot 10
[Sun Dec 04 04:51:09 2005] [notice] jk2_init() Found child 6726 in scoreboard slot 8
[Sun Dec 04 04:51:09 2005] [notice] jk2_init() Found child 6728 in scoreboard slot 6
```

这三行日志能够以一个模式归并：

```
[Sun Dec 04 04:51:* 2005] [notice] jk2_init() Found child 6728 in scoreboard slot *
```

日志按照一定模式归并后，需要被分析的日志量会下降非常多，使得运维工程师很容易从海量日志中洞察日志的价值。

6.2.1 日志模式发现

下面以 Loghub 项目所提供的 Apache 数据集为例，体验日志模式发现功能带来的效果。Loghub 项目地址为【链接 14】。

选择 Apache 的 Acces log 作为样例数据，总数据量有 2000 行：

```
[Sun Dec 04 04:47:44 2005] [notice] workerEnv.init() ok /etc/httpd/conf/workers2.properties
[Sun Dec 04 04:47:44 2005] [error] mod_jk child workerEnv in error state 6
[Sun Dec 04 04:51:08 2005] [notice] jk2_init() Found child 6725 in scoreboard slot 10
[Sun Dec 04 04:51:09 2005] [notice] jk2_init() Found child 6726 in scoreboard slot 8
[Sun Dec 04 04:51:09 2005] [notice] jk2_init() Found child 6728 in scoreboard slot 6
[Sun Dec 04 04:51:14 2005] [notice] workerEnv.init() ok /etc/httpd/conf/workers2.properties
[Sun Dec 04 04:51:14 2005] [notice] workerEnv.init() ok /etc/httpd/conf/workers2.properties
[Sun Dec 04 04:51:14 2005] [notice] workerEnv.init() ok /etc/httpd/conf/workers2.properties
[Sun Dec 04 04:51:18 2005] [error] mod_jk child workerEnv in error state 6
[Sun Dec 04 04:51:18 2005] [error] mod_jk child workerEnv in error state 6
[Sun Dec 04 04:51:18 2005] [error] mod_jk child workerEnv in error state 6
[Sun Dec 04 04:51:37 2005] [notice] jk2_init() Found child 6736 in scoreboard slot 10
[Sun Dec 04 04:51:38 2005] [notice] jk2_init() Found child 6733 in scoreboard slot 7
[Sun Dec 04 04:51:38 2005] [notice] jk2_init() Found child 6734 in scoreboard slot 9
[Sun Dec 04 04:51:52 2005] [notice] workerEnv.init() ok /etc/httpd/conf/workers2.properties
[Sun Dec 04 04:51:52 2005] [notice] workerEnv.init() ok /etc/httpd/conf/workers2.properties
[Sun Dec 04 04:51:55 2005] [error] mod_jk child workerEnv in error state 6
[Sun Dec 04 04:52:04 2005] [notice] jk2_init() Found child 6738 in scoreboard slot 6
[Sun Dec 04 04:52:04 2005] [notice] jk2_init() Found child 6741 in scoreboard slot 9
[Sun Dec 04 04:52:05 2005] [notice] jk2_init() Found child 6740 in scoreboard slot 7
......
```

新建 spell.py，代码实现参考【链接 15】。

编写 lcsobj 对象，主要负责处理最长公共子序列的相关逻辑：

```python
class lcsobj():

    def getlcs(self, seq):
        if isinstance(seq, str) == True:
            seq = re.split(self._refmt, seq.lstrip().rstrip())
        count = 0
        lastmatch = -1
        for i in range(len(self._lcsseq)):
            if self._ispos(i) == True:
                continue
            for j in range(lastmatch+1, len(seq)):
                if self._lcsseq[i] == seq[j]:
```

```python
                    lastmatch = j
                    count += 1
                    break
        return count

    def insert(self, seq, lineid):
        if isinstance(seq, str) == True:
            seq = re.split(self._refmt, seq.lstrip().rstrip())
        self._lineids.append(lineid)
        temp = ""
        lastmatch = -1
        placeholder = False

        for i in range(len(self._lcsseq)):
            if self._ispos(i) == True:
                if not placeholder:
                    temp = temp + "* "
                placeholder = True
                continue
            for j in range(lastmatch+1, len(seq)):
                if self._lcsseq[i] == seq[j]:
                    placeholder = False
                    temp = temp + self._lcsseq[i] + " "
                    lastmatch = j
                    break
                elif not placeholder:
                    temp = temp + "* "
                    placeholder = True
        temp = temp.lstrip().rstrip()
        self._lcsseq = re.split(" ", temp)

        self._pos = self._get_pos()
        self._sep = self._get_sep()

    def param(self, seq):
        if isinstance(seq, str) == True:
            seq = re.split(self._refmt, seq.lstrip().rstrip())
```

```python
            j = 0
            ret = []
            for i in range(len(self._lcsseq)):
                slot = []
                if self._ispos(i) == True:
                    while j < len(seq):
                        if i != len(self._lcsseq)-1 and self._lcsseq[i+1] == seq[j]:
                            break
                        else:
                            slot.append(seq[j])
                        j+=1
                    ret.append(slot)
                elif self._lcsseq[i] != seq[j]:
                    return None
                else:
                    j += 1

            if j != len(seq):
                return None
            else:
                return ret

    def re_param(self, seq):
        if isinstance(seq, list) == True:
            seq = ' '.join(seq)
        seq = seq.lstrip().rstrip()

        ret = []
        print(self._sep)
        print(seq)
        p = re.split(self._sep, seq)
        for i in p:
            if len(i) != 0:
                ret.append(re.split(self._refmt, i.lstrip().rstrip()))
        if len(ret) == len(self._pos):
            return ret
        else:
            return None
```

```python
def _ispos(self, idx):
    for i in self._pos:
        if i == idx:
            return True
    return False

def _tcat(self, seq, s, e):
    sub = ''
    for i in range(s, e + 1):
        sub += seq[i] + " "
    return sub.rstrip()

def _get_sep(self):
    sep_token = []
    s = 0
    e = 0
    for i in range(len(self._lcsseq)):
        if self._ispos(i) == True:
            if s != e:
                sep_token.append(self._tcat(self._lcsseq, s, e))
            s = i + 1
            e = i + 1
        else:
            e = i
        if e == len(self._lcsseq) - 1:
            sep_token.append(self._tcat(self._lcsseq, s, e))
            break

    ret = ""
    for i in range(len(sep_token)):
        if i == len(sep_token)-1:
            ret += sep_token[i]
        else:
            ret += sep_token[i] + '|'
    return ret

def _get_pos(self):
```

```python
            pos = []
            for i in range(len(self._lcsseq)):
                if self._lcsseq[i] == '*':
                    pos.append(i)
            return pos

    def get_id(self):
        return self._id
```

编写 lscmap 类，用于管理 lcsobj 实例，并返回匹配度最高的 lcsobj 对象：

```python
class lcsmap():

    def insert(self, entry):
        seq = re.split(self._refmt, entry.lstrip().rstrip())
        obj = self.match(seq)
        if obj == None:
            self._lineid += 1
            obj = lcsobj(self._id, seq, self._lineid, self._refmt)
            self._lcsobjs.append(obj)
            self._id += 1
        else:
            self._lineid += 1
            obj.insert(seq, self._lineid)

        return obj

    def match(self, seq):
        if isinstance(seq, str) == True:
            seq = re.split(self._refmt, seq.lstrip().rstrip())
        bestmatch = None
        bestmatch_len = 0
        seqlen = len(seq)
        for obj in self._lcsobjs:
            objlen = obj.length()
            if objlen < seqlen/2 or objlen > seqlen*2: continue
```

```
            l = obj.getlcs(seq)
            if l >= seqlen/2 and l > bestmatch_len:
                bestmatch = obj
                bestmatch_len = l
        return bestmatch

    def objat(self, idx):
        return self._lcsobjs[idx]

    def size(self):
        return len(self._lcsobjs)
```

编写主函数 main.py：

```
import spell as s

slm = s.lcsmap('[\\s]+')
with open('./Apache_2k.log','r') as f:
    lines=f.readlines()
    for row in lines:
        slm.insert(row)
```

执行上述代码后，算法会对所读取的日志进行分析，得到一个日志模式的模型：

```
with open('./Apache_2k.log','r') as f:
    lines=f.readlines()
    for row in lines:
        print(' '.join(slm.match(row)._lcsseq))
```

对 2000 条 Apache Access Log 进行模式发现，得到如下结果：

```
* Dec * 2005] [notice] workerEnv.init() ok /etc/httpd/conf/workers2.properties
[Sun Dec 04 * 2005] [error] mod_jk child *
[Sun Dec 04 * 2005] * jk2_init() * child * in scoreboard *
[Sun Dec 04 * 2005] * jk2_init() * child * in scoreboard *
[Sun Dec 04 * 2005] * jk2_init() * child * in scoreboard *
```

```
* Dec * 2005] [notice] workerEnv.init() ok /etc/httpd/conf/workers2.properties
* Dec * 2005] [notice] workerEnv.init() ok /etc/httpd/conf/workers2.properties
* Dec * 2005] [notice] workerEnv.init() ok /etc/httpd/conf/workers2.properties
[Sun Dec 04 * 2005] [error] mod_jk child *
[Sun Dec 04 * 2005] [error] mod_jk child *
[Sun Dec 04 * 2005] [error] mod_jk child *
[Sun Dec 04 * 2005] * jk2_init() * child * in scoreboard *
[Sun Dec 04 * 2005] * jk2_init() * child * in scoreboard *
[Sun Dec 04 * 2005] * jk2_init() * child * in scoreboard *
* Dec * 2005] [notice] workerEnv.init() ok /etc/httpd/conf/workers2.properties
* Dec * 2005] [notice] workerEnv.init() ok /etc/httpd/conf/workers2.properties
……
```

可以看到，Spell 算法已经将变动的文本打上了星号，保留了没有变动的文本。

使用 Pandas 对数据进行分组：

```
df=pd.DataFrame(df_list)
df.value_counts()
```

2000 条 Apache Access Log 最终变成了 7 组，第一列是这一组的模式，第二列是这一组的总数量，在海量日志的分析下，使用模式发现的效果会更加明显。

```
* Dec * 2005] [notice] workerEnv.init() ok /etc/httpd/conf/workers2.properties      569
[Sun Dec 04 * 2005] * jk2_init() * child * in scoreboard *                          448
[Mon Dec 05 * 2005] * jk2_init() * child * in scoreboard *                          400
[Sun Dec 04 * 2005] [error] mod_jk child *                                          287
[Mon Dec 05 * 2005] * child * in * 6                                                180
[Mon Dec 05 * 2005] [error] mod_jk child *                                           84
* Dec * 2005] [error] [client * Directory index forbidden by rule: /var/www/html/    32
```

6.2.2 日志模式统计分析

有了日志模式发现模型，我们就可以对每天的日志做一个模式发现，得到每天的分组模式分布情况。假设有以下两天的模式发现结果。

第一天：

```
* Dec * 2005] [notice] workerEnv.init() ok /etc/httpd/conf/workers2.properties        569
[Sun Dec 04 * 2005] * jk2_init() * child * in scoreboard *                             448
[Mon Dec 05 * 2005] * jk2_init() * child * in scoreboard *                             400
[Sun Dec 04 * 2005] [error] mod_jk child *                                             287
[Mon Dec 05 * 2005] * child * in * 6                                                   180
[Mon Dec 05 * 2005] [error] mod_jk child *                                              84
```

第二天：

```
* Dec * 2005] [notice] workerEnv.init() ok /etc/httpd/conf/workers2.properties       1569
[Sun Dec 04 * 2005] * jk2_init() * child * in scoreboard *                             448
[Mon Dec 05 * 2005] * jk2_init() * child * in scoreboard *                             400
[Sun Dec 04 * 2005] [error] mod_jk child *                                             287
[Mon Dec 05 * 2005] * child * in * 6                                                   180
[Mon Dec 05 * 2005] [error] mod_jk child *                                              84
* Dec * 2005] [error] [client * Directory index forbidden by rule: /var/www/html/       10
```

通过对两天的日志模式进行统计分析，可以快速得出两个结论：

（1）* Dec * 2005] [notice] workerEnv.init() ok /etc/httpd/conf/workers2.properties 模式的日志增长率过高，和前一天的模式数量差异过大，需要引起关注。

（2）* Dec * 2005] [error] [client * Directory index forbidden by rule: /var/www/html/模式在第一天没有出现，也属于需要引起关注的日志。

可以看到，通过日志模式统计分析，能够快速在海量日志中得到一个分析的方向，在没有日志模式发现功能之前，这是不可想象的。

6.2.3 实时异常检测

既然能够对日志进行批处理的日志模式发现，自然也能够使用 Spell 算法对日志进行实时流式的异常检测。使用 Spell 算法训练后的模型能够被导出为模型文件，算法服务加载此模型文件，在 Kafka 中对日志进行消费。当发现异常的日志时，给日志打上异常标签，系统就可以根据此标签触发告警或者生成事件了。实时异常检测的设计方案如图 6-7 所示。

采用日志模式发现之后，很多事情都变得容易了。例如，我们能够实时地发现异常日志，将告警内容进行日志模式发现，按照告警模式对告警进行分组，对日志模式进行实时的统计分析，等等。日志模型发现工具 Spell 是 AIOps 工具包里面的一件非常实用的工具。

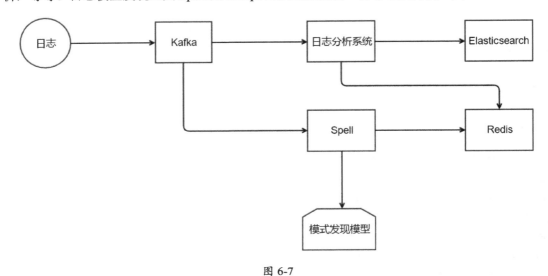

图 6-7

6.3 告警关联分析

在日常运维过程中，我们经常需要判断某个指标的异常是否由其他指标的异常而引起的。一个典型的案例是 Web 拨测发现系统已经无响应了，系统给运维工程师发出了告警信息，根据运维工程师的经验，找到 CPU、内存、I/O 指标的曲线，然后分析究竟是哪个指标的异常导致应用系统无响应的。这在应用架构比较简单的时候是可行的，因为服务器数量、应用数量，以及所监控的指标数量都比较小，运维工程师凭借自身的经验还能够处理得过来。但是，一方面随着监控系统的完善，现在运维工程师所能掌握的指标非常多，光服务器层面的监控指标就轻易超过了 100 个。另一方面，随着容器、微服务等技术的普遍应用，运维工程师需要面对的应用拓扑也从以前的单体应用变成了分布式应用，快速找到关键的指标能够有效地提升故障处理的效率。

微软在 2014 年 SIGKDD 会议上发表了一篇论文 *Correlating Events with Time Series for Incident Diagnosis*，这篇论文提出了一种无监督和统计判别的算法，可以检测出事件（E）与时间序列（S）的关联关系，并且可以检测出时间序列（S）的单调性（上升或者下降）。利用这个算法，可以做到当出现告警后，将与此告警相关的指标找出来，提升告警分析的效率，可以将告警关联分析的能力（【链接 16】）加入算法工具包。

下面我们看一下使用此工具能够达到怎样的效果？示例数据所使用的指标如表 6-1 所示。

表 6-1

指标项	中文含义	功能
host.alive	存活监控	报警项
cpu.idle	CPU 空闲率	监控项
mem.swapused.percent	Swap 使用率	监控项
mem.memused.percent	内存使用率	监控项
net.if.total.bits.sum	网卡流量	监控项
ss.closed	closed 状态连接数	监控项

host.alive 是告警项，cpu.idle、mem.swapused.percent、mem.memused.percent、net.if.total.bits.sum 及 ss.closed 是监控项，需要解决的问题是分析哪些监控项与 host.alive 告警是有关联的。整个项目包含一份关键的代码。执行 PYTHONPATH=pwd python3 ./examples/alarm_association.py，会得到如下结果：

```
cpu.idle is not related to alarm
net.if.total.bits.sum is related to alarm
mem.memused.percent is related to alarm
mem.swapused.percent is related to alarm
ss.closed is not related to alarm
```

可以看到，算法会发现与 host.alive 告警相关的指标有 net.if.total.bits.sum、mem.memused.percent、mem.swapused.percent。

它是如何实现的？在 *Correlating Events with Time Series for Incident Diagnosis* 论文中，告警关联分析分解为三个问题：

（1）事件和时间序列是否存在关联性。

（2）关联关系的因果关系，是事件导致了时间序列的变化，还是时间序列导致了事件的发生。

（3）关联关系的单调性影响，用于判断时间序列是突增还是突降了。

对应到代码的实现，可以参考 aiopstools 项目中的 alarm_association.py：

```
from __future__ import division
import random
import math
from operator import import itemgetter
```

```python
def mixdata(alarmtime, timeseries_set, timeseries):
    """
    :param alarmtime: 报警的时刻序列，已经排好顺序
    :param timeseries_set: 每个报警时刻的时间序列构成的时序集
    :param timeseries: 报警时刻整体区间内的时序数据
    :return:mixset 是混合集，alarm_number 是报警样本个数，random_number 是随机样本个数
    """
    mixset = []
    alarm_number = 0
    random_number = 0
    timegap = int((alarmtime[-1]-alarmtime[0]) / 600)
    randomnum = min(len(alarmtime), int(3*timegap/4)) - 1
    for i in range(len(alarmtime)):
        data = timeseries_set[i]
        data.append('alarm')
        if len(data) > 1 and data not in mixset:
            mixset.append(data)
            alarm_number += 1
    while randomnum > 0:
        end = random.randint(5, len(timeseries))
        start = end - 5
        data = timeseries[start:end]
        data.append("random")
        randomnum -= 1
        if len(data) > 1 and data not in mixset:
            mixset.append(data)
            random_number += 1
    return mixset, alarm_number, random_number


def distance(data1,data2):
    dis = 0
    for i in range(0, len(data1)-1):
        dis += (data1[i]-data2[i]) ** 2
    dis = math.sqrt(dis)
    return dis
```

```python
def feature_screen(mixset, alarm_number, random_number):
    """
    :param mixset: 报警序列与随机序列的混合集
    :param alarm_number: 报警序列个数
    :param random_number: 随机序列个数
    :return: 监控项与报警是否相关
    """
    if alarm_number == 0 or random_number == 0:
        return False
    sum_number = alarm_number + random_number
    mean = (alarm_number/sum_number) ** 2 + (random_number/sum_number) ** 2
    stdDev = (alarm_number/sum_number) * (random_number/sum_number) * (1 + 4 *
(random_number/sum_number) * (alarm_number / sum_number))
    R = 10
    trp = 0
    alapha = 1.96
    for j in range(len(mixset)):
        tempdic = {}
        for k in range(len(mixset)):
            if j == k:
                continue
            dis = distance(mixset[j], mixset[k])
            tempdic.setdefault(k, dis)
        temp_list = sorted(tempdic.items(), key=itemgetter(1), reverse=False)[0:R]
        for k in temp_list:
            if mixset[j][-1] == mixset[k[0]][-1]:
                trp += 1

    trp = float(trp / (R*sum_number))
    check = (abs(trp-mean) / stdDev) * math.sqrt(R*sum_number)
    if check > alapha:
        return True
    return False


def get_GR(alarmseries,nomalseries):
    '''
    :param alarmseries: 单一报警的时间序列
```

```
:param nomalseries: 整体报警的时间序列
:return:
'''
cutnum = 10   # 切分份数
maxvalue = float("-inf")
minvalue = float("inf")
GR = 0
while None in alarmseries:
    alarmseries.remove(None)
C1 = len(alarmseries)
if max(alarmseries) > maxvalue:
    maxvalue = max(alarmseries)
if min(alarmseries) < minvalue:
    minvalue = min(alarmseries)
while None in nomalseries:
    nomalseries.remove(None)
C2 = len(nomalseries)
if max(nomalseries) > maxvalue:
    maxvalue = max(nomalseries)
if min(nomalseries) < minvalue:
    minvalue = min(nomalseries)
value_gap = (maxvalue-minvalue) / cutnum
print(C1)
print(C2)
if C1 == 0 or C2 == 0 or value_gap == 0:
    return GR
HD = (C1 / (C1+C2)) * math.log((C1 / (C1+C2)), 2) + (C2 / (C1+C2)) * math.log((C2 / (C1+C2)), 2)
Neg = [0] * (cutnum+1)
Pos = [0] * (cutnum+1)
for value in alarmseries:
    temp_count = int((value-minvalue) / value_gap) + 1
    if temp_count > cutnum:
        temp_count = cutnum
    Neg[temp_count] += 1
for value in nomalseries:
    temp_count = int((value-minvalue) / value_gap) + 1
    if temp_count > cutnum:
```

```
            temp_count = cutnum
        Pos[temp_count] += 1
    HDA = 0
    HAD = 0
    for j in range(1, cutnum + 1):
        temp = 0
        if Neg[j] != 0 and Pos[j] != 0:
            HAD += ((Neg[j]+Pos[j]) / (C1+C2)) * math.log(((Neg[j]+Pos[j]) / (C1+C2)), 2)
            temp = (Neg[j] / (Neg[j]+Pos[j])) * math.log((Neg[j] / (Neg[j]+Pos[j])), 2)
+ (Pos[j] / (Neg[j]+Pos[j])) * math.log((Pos[j] / (Neg[j]+Pos[j])), 2)
        elif Neg[j] == 0 and Pos[j] != 0:
            HAD += ((Neg[j]+Pos[j]) / (C1+C2)) * math.log(((Neg[j]+Pos[j]) / (C1 + C2)), 2)
        elif Pos[j] == 0 and Neg[j] != 0:
            HAD += ((Neg[j]+Pos[j]) / (C1+C2)) * math.log(((Neg[j]+Pos[j]) / (C1+C2)), 2)
        HDA += ((Neg[j]+Pos[j]) / (C1+C2)) * temp
    GR = (HD - HDA) / HAD
    return GR
```

有了此工具包之后，我们可以将它嵌入告警引擎，当发生告警之后，将指标与告警一起作为此工具包的输入，得到输出后，在告警中附上相关的建议，缩小运维工程师需要排查的指标的范围，提升故障排查的效率。值得注意的是，告警发送和告警关联分析所给出的建议不要串联实现，因为在生产环境中，由于指标数量繁多，计算需要花费一点时间，而告警是需要及时送达给运维工程师的，建议在分析完告警关联性之后，采用异步的方式写入告警记录。

告警管理分析落地模式如图 6-8 所示。

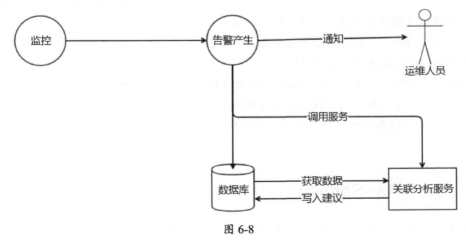

图 6-8

告警关联分析是 AIOps 的一个常见场景，也是非常容易提升效率的场景，建议将此工具加入读者的 AIOps 工具包。

6.4 语义检索

语义检索功能常用在知识库模块上，在日常运维过程中，运维工程师往往积累了大量的知识，但是又没有一个便捷的工具能够找到这些知识。Elasticsearch 这样的数据库能够通过倒排索引帮助运维工程师通过关键字对数据进行检索，使用倒排索引进行检索的优点是准确度高，缺点是无法根据检索者所表达的语义进行检索，将最合适的知识推荐给使用者。倒排索引检索如图 6-9 所示。

图 6-9

相比倒排索引检索的模式，语义检索的模式是将目标检索语句编码为一个矩阵，再通过矩阵在数据库中寻找与此矩阵相似度最高的语句，返回语义最相近的检索结果。语义检索如图 6-10 所示。

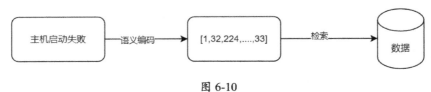

图 6-10

在预训练语义模型出现之前，要完成语义检索是一件相对麻烦的事情，一方面需要准备大量的语料作为训练素材，另一方面需要有强大的算力才能完成语言模型的训练。当预训练语义模型出现之后，要完成语义检索任务就变得简单多了。其中最有代表性的为 Bert 系列，谷歌的 Bert 一经推出就打破了十多项纪录，NLP 领域语义检索的精确度也随着 Bert 的出现得到了不少的提升。在 AIOps 上，我们也可以使用 Bert 对知识库的知识进行语义编码，完成知识语义检索的功能。通过语义检索搭配倒排索引检索的功能，为运维工程师提供更加强大的运维知识库能力。

6.4.1 Bert-As-Service

Bert-As-Service 项目是一个将 Bert 的功能封装好的"开箱即用"的工具包，通过使用此工具包，我们能够以极低的成本完成语义检索的功能。

安装 Bert-As-Service：

```
pip3 install bert-serving-server
pip3 install bert-serving-client
pip3 install tensorflow-gpu==1.13.4
```

Bert-As-Service 安装完毕后需要使用 Bert 的预训练模型，这里我们使用哈尔滨工业大学提供的 Bert 预训练模型（【链接 17】），下载"BERT-wwm, Chinese"预训练模型。

模型下载完毕后，启动服务端：

```
bert-serving-start -model_dir /tmp/english_L-12_H-768_A-12/ -num_worker=1
```

```
I:GRAPHOPT:[gra:opt:132]:load parameters from checkpoint...
I:GRAPHOPT:[gra:opt:136]:optimize...
I:GRAPHOPT:[gra:opt:144]:freeze...
I:GRAPHOPT:[gra:opt:149]:write graph to a tmp file: /tmp/tmp05ieqgst
I:VENTILATOR:[__i:_i: 75]:optimized graph is stored at: /tmp/tmp05ieqgst
I:VENTILATOR:[__i:_ru:129]:bind all sockets
I:VENTILATOR:[__i:_ru:133]:open 8 ventilator-worker sockets
I:VENTILATOR:[__i:_ru:136]:start the sink
I:SINK:[__i:_ru:306]:ready
I:VENTILATOR:[__i:_ge:222]:get devices
I:VENTILATOR:[__i:_ge:255]:device map:
                    worker  0 -> gpu  0
I:WORKER-0:[__i:_ru:531]:use device gpu: 0, load graph from /tmp/tmp05ieqgst
I:WORKER-0:[__i:gen:559]:ready and listening!
I:VENTILATOR:[__i:_ru:164]:all set, ready to serve request!
```

当看到"I:VENTILATOR:[__i:_ru:164]:all set, ready to serve request!"的时候，表示服务端启动完成，接下来可以使用客户端 SDK 获取文本的语义编码：

```
from bert_serving.client import BertClient
bc = BertClient()
rs=bc.encode(['IP 为 127.0.0.1 的主机启动失败','Zabbix 告警:主机 CPU 负载过高'])
print(rs)
```

将我们期望被语义编码的知识标题传入 encode 函数，得到语义编码矩阵：

```
[[ 0.58720016  0.06937832  0.6031635  ...  0.5306102  -0.06607438
  -0.39411104]
 [ 0.39919484 -0.0757445   0.16896145 ...  0.1066585  -0.49570328
  -0.18832839]]
```

接下来尝试一下语义检索，对示例代码进行一些修改：

```
from bert_serving.client import BertClient
import numpy as np
bc = BertClient()
questions=['主机启动失败','Zabbix 告警:主机 CPU 负载过高']
doc_vecs=bc.encode(questions)

query_vec = bc.encode(['系统无法开机'])[0]
score = np.sum(query_vec * doc_vecs, axis=1)
topk_idx = np.argsort(score)[::-1]
for idx in topk_idx:
    print('> %s\t%s' % (score[idx], questions[idx]))
```

上面的代码对"主机启动失败"和"Zabbix 告警:主机 CPU 负载过高"两个标题进行了语义编码，然后输入检索的标题"系统无法开机"进行检索。假设我们只用倒排索引引擎进行运维知识管理，输入"系统无法开机"进行检索无法检索到任何内容，但是搭配语义编码，能够得到如下结果：

```
> 401.58398    主机启动失败
> 392.65845    Zabbix告警:主机CPU负载过高
```

可以看到"主机启动失败"和"系统无法开机"在语义上更相近，此时系统就可以将"主机启动失败"的知识推荐给使用者了。由于语义检索并不是精准匹配，所以在系统设计上，建议使用推荐的模式，将语义上最匹配的几个标题推送给运维工程师，通过推荐的模式优化使用体验。

6.4.2　Bert Fine-tuning

语义检索的准确率很大程度取决于 Bert-As-Service 所加载的模型，6.4.1 节中使用的是哈工大提供的已经训练好的语言模型，在运维领域有一些专有词汇，或者工作中的特定短语，我们期望将它们加入预训练语言模型，而不是从头进行语言模型训练，这时就需要使用 Bert Fine-tuning 技术了。

首先，复制 Bert 的代码(【链接 18】)，然后准备一下数据，创建一个文件夹，名字叫作 data_dir，在里面分别创建 dev.tsv、test.tsv、train.tsv。三份 tsv 文件的数据格式如下：

```
"分类标签"   "数据"
```

接着在 run_classifier.py 中新增一些代码。先添加一个自己的处理任务，这里我们把任务名

字叫作 sample,指定 Processor:

```python
processors = {
    "cola": ColaProcessor,
    "mnli": MnliProcessor,
    "mrpc": MrpcProcessor,
    "xnli": XnliProcessor,
    'sample': SampleProcessor,
}
```

编写数据处理部分的代码:

```python
class SampleProcessor(DataProcessor):
    def __init__(self):
        self.language = 'zh'

    def load_sample_data(self, path, guid_prefix):
        file_path = Path(path)
        with open(file_path, 'r', encoding='utf-8') as file:
            reader = file.readlines()
        examples = []
        for index, line in enumerate(reader):
            guid = '%s-%d' % (guid_prefix, index)
            split_line = line.strip().split('\t')
            text_a = tokenization.convert_to_unicode(split_line[1])
            label = split_line[0]
            examples.append(InputExample(guid=guid, text_a=text_a,
                                         text_b=None, label=label))
        return examples

    def get_train_examples(self, data_dir):
        return self.load_sample_data('./data_dir/train.tsv', 'train')

    def get_labels(self):
        label = []
        for x in range(1, 14):
            label.append('"%s"' % x)
        return label
```

```python
    def get_dev_examples(self, data_dir):
        return self.load_sample_data('./data_dir/dev.tsv', 'dev')

    def get_test_examples(self, data_dir):
        return self.load_sample_data('./data_dir/test.tsv', 'test')
```

上述代码把数据集进行三路划分（按照训练集、开发集、测试集进行划分），根据数据集的格式，把类别和数据提取出来。text_a 和 text_b 是 Bert 训练出来的产物，其中的一个功能就是推断句子 A 的下一句是不是句子 B，所以只有在做句子对训练的时候才会用得上。

一切准备就绪，可以开始对模型做"模型微调"了。

```
python3 ./run_classifier.py \
    --task_name=sample \
    -do_eval=true \
    --do_predict=true \
    --data_dir=/root/bert/data_dir/ \
    --bert_config_file=/root/publish/bert_config.json \
    --vocab_file=/root/publish/vocab.txt \
    --init_checkpoint=/root/models/ \
    --max_seq_length=128 \
    --output_dir=/tmp/results
```

执行"模型微调"后，会输出"模型微调"后的精度：

```
eval_accuracy = 0.6507937
eval_loss = 1.2142951
global_step = 0
loss = 1.209952
```

训练完成后，就可以使用 6.4.1 节中的方法加载并使用自己训练好的模型了。

6.5 异常检测

异常检测是运维中必不可少的工具之一了，也是比较成熟的 AIOps 领域之一，异常检测应用的场景也非常广泛，如应用行为异常检测、网络流量异常检测、监控指标异常检测。

6.5.1 典型场景——监控指标异常检测

运维工程师每天都要查看非常多的指标,在监控、告警、分析各个环节,指标都伴随着运维工程师,这对运维工程师来说,信息量是过载的。我们可以把异常检测的功能嵌入指标查看页面,对监控指标做异常检测,然后将异常点在监控面板中标注出来,经过标注后,运维工程师只需要查看少量的监控点,极大地提升了运维效率。同样地,在系统后台形成异常检测报表,以邮件形式发送给运维负责人,也是一种提升运维效率的不错方法。CPU 使用率监控指标如图 6-11 所示。

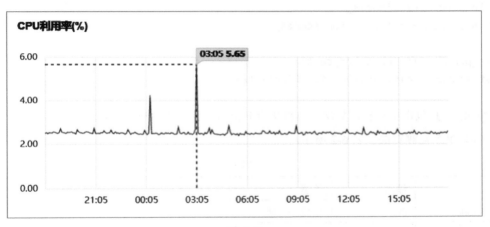

图 6-11

6.5.2 异常检测工具包——PyOD

有非常多的 AIOps 工具包可用于异常检测,例如,PyOD 就提供了数十种异常检测的工具包,PyOD 提供了一致的、易用的 API,我们直接将数据传入即可,非常简单易用。下面以一个实例介绍 PyOD 的使用方式。

首先安装 PyOD:

```
pip install pyod
```

然后使用 IForest 算法进行异常检测。其中:

- clf.labels_——返回训练数据上的分类标签(0:正常值;1:异常值);
- clf.decision_scores_——返回训练数据上的异常值(分值越大越异常);

- predict 函数——返回未知数据上的分类标签（0：正常值；1：异常值）；
- decision_function 函数——返回未知数据上的异常值（分值越大越异常）。

使用 PyOD 做异常检测的示例如下：

```
from pyod.models.iforest import IForest

clf = IForest()
clf.fit(X_train)

y_train_pred = clf.labels_
y_train_scores = clf.decision_scores_

y_test_pred = clf.predict(X_test)
y_test_scores = clf.decision_function(X_test)
```

PyOD 工具包由单个检测算法、离群集合和离群检测器组合框架组成。

（1）单个检测算法如表 6-2 所示。

表 6-2

类型	缩写	简述	年份
Linear Model	PCA	主成分分析	2003
Linear Model	MCD	最小协方差行列式	1999
Linear Model	OCSVM	One-Class Support Vector Machines	2001
Linear Model	LMDD	Deviation-based 异常值检测（LMDD）	1996
Proximity-Based	LOF	Local Outlier Factor	2000
Proximity-Based	COF	Connectivity-Based Outlier Factor	2002
Proximity-Based	CBLOF	Clustering-Based Local Outlier Factor	2003
Proximity-Based	LOCI	位点：基于局部相关积分的快速离群点检测	2003
Proximity-Based	HBOS	Histogram-based Outlier Score	2012
Proximity-Based	kNN	k 个最近邻（使用第 k 个最近邻的距离作为离群值）	2000
Proximity-Based	AvgKNN	平均 kNN（使用 k 个最近邻的平均距离作为离群值）	2002
Proximity-Based	MedKNN	中值 kNN（使用 k 个最近邻的中间距离作为异常值得分）	2002
Proximity-Based	SOD	Subspace Outlier Detection	2009
Probabilistic	ABOD	Angle-Based Outlier Detection	2008

续表

类型	缩写	简述	年份
Probabilistic	FastABOD	基于近似的快速 Angle-Based 离群点检测	2008
Probabilistic	SOS	Stochastic Outlier Selection	2012
Outlier Ensembles	IForest	Isolation Forest	2008
Outlier Ensembles	—	Feature Bagging	2005
Outlier Ensembles	LSCP	LSCP：并行离群点集合的局部选择性组合	2019
Outlier Ensembles	XGBOD	基于极值 Boosting 的离群点检测	2018
Outlier Ensembles	LODA	轻型 On-line 异常探测器	2016
Neural Networks	AutoEncoder	全连接自动编码器	2015
Neural Networks	VAE	变分自动编码器	2013
Neural Networks	Beta-VAE	可变自动编码器	2018
Neural Networks	SO_GAAL	Single-Objective Generative Adversarial Active Learning	2019
Neural Networks	MO_GAAL	Multiple-Objective Generative Adversarial Active Learning	2019

（2）离群集合和离群检测器组合框架如表 6-3 所示。

表 6-3

类型	缩写	简述	年份
Outlier Ensembles	—	Feature Bagging	2005
Outlier Ensembles	LSCP	LSCP：并行离群点集合的局部选择性组合	2019
Outlier Ensembles	XGBOD	基于极值 Boosting 的离群点检测（监督）	2018
Outlier Ensembles	LODA	轻型 On-line 异常探测器	2016
Combination	Average	平均得分的简单组合	2015
Combination	Weighted Average	检测器加权平均得分的简单组合	2015
Combination	Maximization	取最大分数的简单组合	2015
Combination	AOM	最大值平均值	2015
Combination	MOA	平均值最大化	2015
Combination	Median	取分数中位数的简单组合	2015
Combination	majority Vote	通过获得标签多数票的简单组合（可使用权重）	2015

PyOD 提供了包括 IForest、LOF、COF 等数十种异常检测方法，非常推荐加入 AIOps 工具包。

6.6 时序预测

时序预测是运维领域的老话题了，和异常检测的使用频率一样高，常用于告警的动态阈值、存储容量分析等功能，适合为 APM、NPM、基础监控等监控工具提供 AIOps 的能力。

6.6.1 典型场景——动态告警阈值

以笔者的云环境为例，笔者在云平台上购买了一台包年的云主机（200Mbps 的按量付费带宽），由于是按流量计费，每 GB 流量的价格是 0.8 元，最近 7 天的网络流量带宽使用情况如图 6-12 所示。

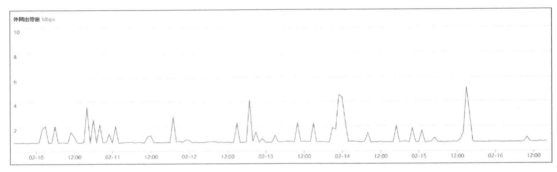

图 6-12

对于用户来说，是非常关注流量的使用率的。此时，就应该对流量的使用情况进行告警。在不使用 AIOps 工具的时候，我们通常会对流量进行固定阈值告警。例如，带宽使用率大于 95% 时发出告警。这种配置方式看似合理，但实际上几乎没有任何作用，因为用户是无法知道究竟阈值配置为多少是合理的。而此时，结合时序预测的功能就能非常好地解决阈值配置的问题。首先，我们依然要配置固定阈值，流量使用率大于 95%时就告警。然后，系统通过将用户的流量使用率的历史数据交给时序预测工具训练后，得到时序预测模型，通过时序预测模型预测未来的指标运行区间带，以此区间带作为动态告警阈值，超过了此区间带则视为异常，将告警信息发送给用户。

6.6.2 时序预测工具包——Prophet

对于时序预测，比较主流的观点认为其受四种成分影响。

- 趋势：宏观、长期、持续性的作用力；
- 周期：比如商品价格在较短的时间内围绕某个均值上下波动；

- 季节：变化规律相对固定，并呈现某种周期特征；"季节"不一定按年计；每周、每天的不同时段的规律，也可称作季节性；
- 随机：随机的不确定性。

这四种成分对时间序列的影响常归纳为累积和相乘两种。累积意味着四种成分相互叠加，它们之间相对独立，相互影响较小。而相乘意味着它们相互影响更为明显。在时序预测的发展历史中，从 AR、MA、ARMA、ARIMA、SARIMA 一路演变，但是 SARIMA 的使用依然比较麻烦，所以 Facebook 推出了一个能够兼顾使用方便和预测质量的工具包 Prophet。

相比目前其他时序预测工具，Prophet 主要有以下两点优势。

- 使用时序预测变得非常容易，默认情况下不需要配置任何参数，即可直接训练模型并得到质量较高的预测结果；
- 它是为非专家"量身定制"的，可以直接通过修改季节参数来拟合季节性，修改趋势参数来拟合趋势信息，指定假期来拟合假期信息，等等。没有复杂的参数，调整起来非常便捷。

Prophet 遵循 sklearn 模型 API。我们可以创建一个 Prophet 类的实例，然后调用它的 fit 和 predict 方法。Prophet 的输入必须包含 ds 和 y 两列数据，其中 ds 是时间戳列，必须是时间信息；y 列必须是数值，代表我们需要预测的信息。下面使用 Prophet 对时序数据进行预测：

```
import pandas as pd
from fbprophet import Prophet

df = pd.read_csv('data.csv')
m = Prophet()
m.fit(df)

future = m.make_future_dataframe(periods=365)
forecast = m.predict(future)

m.plot(forecast)
m.plot_components(forecast)
```

图 6-13 中的点是实际发生的指标点，区间带是使用 Prophet 预测后，由上限和下限组成的区间带。图 6-14 分别展示了趋势和周期性分量的情况。

在时序预测领域有非常多的模型和方法可供我们使用，比如 LSTM、SARIMA 等，但是 Prophet 具备"开箱即用"、开销较 LSTM 等模型较小等特点，可以将其纳入我们的 AIOps 工具包。

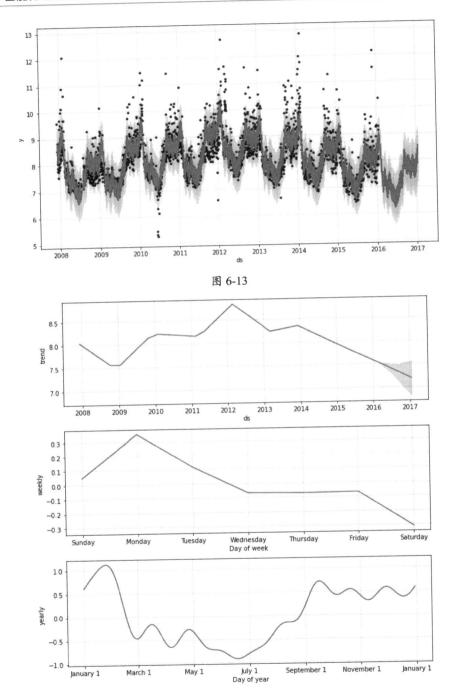

图 6-13

图 6-14

第 7 章
加速 AIOps 落地——
AI 平台

7.1 AI 平台与 AIOps

7.1.1 为运维系统插上 AI 的翅膀

实现较规范的 AI 运维功能至少需要经过数据管理、数据标注、模型管理、训练任务、自动调参等过程。除了在业务层面上能看到的模型生产的流程，由于 AI 模型训练过程中存在样本数据量较大、有显卡资源使用需求等特点，平台还需要对 AI 模型的训练，以及推理过程中所使用的计算、存储、网络等资源进行综合有效的管理。

这些特点要直接嵌入现有的运维系统中，改造的工作量是非常大的，一方面需要对现有运维系统的业务系统流程进行改造，另一方面需要引入非常多让 AI 模型能够被可靠、标准化生产出来的能力。

AI 平台的出现成为 AIOps 标准化落地的一个可能的方向，AI 平台本身是为了让 AI 能力能够快速生产而诞生的。以目标检测为例，生产一个目标检测模型最少需要经过对样本图片的统一管理、样本标注、模型训练等环节，最终才能将目标检测的 AI 能力标准化地提供给 AI 能力使用者。

同样，我们可以把 AI 平台和现有的运维系统进行能力整合，形成一套较为标准化的 AIOps 搭建方案，如图 7-1 所示。

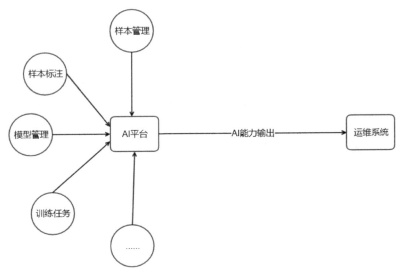

图 7-1

通过标准化的 AI 能力输出方式，可以让 AI 平台为运维系统提供 AI 能力，运维系统既可以使用 AI 平台能力构建自身的 AIOps 场景，又可以实现 AI 平台和运维系统的隔离。算法开发人员在 AI 平台上进行 AI 能力的开发，通过 AI 平台的资源管理调度能力对模型进行训练，形成模型后通过 RESTful API 或 RPC 的方式为运维系统提供 AI 能力。运维系统负责人通过对自身的运维系统进行微小的改造，在需要 AI 能力的位置引入 AI 平台的能力，达到 AIOps 快速落地的目的。

7.1.2　Polyaxon

Polyaxon 是一个用 Python 编写的生产级的 AI 平台，使用 Polyaxon，能够让 AI 模型的生产更加流程化和标准化，它具有以下优点：

- 支持 TensorFlow、Keras、Torch、Caffe 等主流 AI 框架；
- 提供在线开发的能力，AI 模型的生产者可以直接在 Polyaxon 上进行 AI 模型的开发；
- 具备企业级系统所需要的安全、权限、分析等能力；
- 提供了一套完整的 AI 模型开发工具。

可以通过 docker-compose 的方式，快速地部署一套 Polyaxon 进行体验。新建 base.env 配置文件，此文件声明了 Polyaxon 部署过程中所使用的基础配置：

```
POLYAXON_K8S_NAMESPACE=polyaxon
POLYAXON_K8S_NODE_NAME=compose
POLYAXON_K8S_APP_NAME=polyaxon-compose
POLYAXON_ENVIRONMENT=compose
POLYAXON_ENABLE_SCHEDULER=0
POLYAXON_CHART_IS_UPGRADE=0
POLYAXON_REDIS_HEARTBEAT_URL=redis://redis:6379/8
POLYAXON_REDIS_GROUP_CHECKS_URL=redis://redis:6379/9
POLYAXON_HEARTBEAT_URL=redis://redis:6379/8
POLYAXON_GROUP_CHECKS_URL=redis://redis:6379/9
POLYAXON_K8S_GPU_RESOURCE_KEY=""
POLYAXON_DIRS_NVIDIA={}
POLYAXON_K8S_APP_CONFIG_NAME=""
POLYAXON_K8S_APP_SECRET_NAME=""
POLYAXON_K8S_RABBITMQ_SECRET_NAME=""
POLYAXON_K8S_DB_SECRET_NAME=""
POLYAXON_PERSISTENCE_DATA={"data": {"mountPath": ""}}
POLYAXON_PERSISTENCE_LOGS={"mountPath": ""}
POLYAXON_PERSISTENCE_OUTPUTS={"outputs": {"mountPath": ""}}
POLYAXON_PERSISTENCE_REPOS={"mountPath": ""}
POLYAXON_PERSISTENCE_UPLOAD={"mountPath": ""}
POLYAXON_K8S_SERVICE_ACCOUNT_NAME=""
POLYAXON_K8S_RBAC_ENABLED=0
POLYAXON_K8S_INGRESS_ENABLED=0
POLYAXON_ROLE_LABELS_WORKER=polyaxon-workers
POLYAXON_ROLE_LABELS_DASHBOARD=polyaxon-dashboard
POLYAXON_ROLE_LABELS_LOG=polyaxon-logs
POLYAXON_ROLE_LABELS_API=polyaxon-api
POLYAXON_TYPE_LABELS_CORE=polyaxon-core
POLYAXON_TYPE_LABELS_RUNNER=polyaxon-runner
POLYAXON_ROLE_LABELS_CONFIG=polyaxon-config
POLYAXON_ROLE_LABELS_HOOKS=polyaxon-hooks
POLYAXON_K8S_API_HOST=localhost
POLYAXON_K8S_API_HTTP_PORT=8000
POLYAXON_K8S_API_WS_PORT=1337
POLYAXON_CHART_VERSION=0.5.1
```

```
POLYAXON_CLI_MIN_VERSION=0.5.0
POLYAXON_CLI_LATEST_VERSION=0.5.1
POLYAXON_PLATFORM_MIN_VERSION=0.5.0
POLYAXON_PLATFORM_LATEST_VERSION=0.5.1
```

新建 components.env 文件，此文件声明了 Polyaxon 启动过程中的连接信息，如 Redis、DB 等连接配置：

```
POLYAXON_DB_NAME=polyaxon
POLYAXON_DB_USER=polyaxon
POLYAXON_DB_PASSWORD=polyaxon
POLYAXON_DB_HOST=postgres
POLYAXON_DB_PORT=5432
POLYAXON_BROKER_BACKEND=redis
POLYAXON_REDIS_CELERY_BROKER_URL=redis:6379/0
POLYAXON_REDIS_CELERY_RESULT_BACKEND_URL=redis:6379/1
POLYAXON_REDIS_JOB_CONTAINERS_URL=redis:6379/3
POLYAXON_REDIS_TO_STREAM_URL=redis:6379/4
POLYAXON_REDIS_SESSIONS_URL=redis:6379/5
POLYAXON_REDIS_EPHEMERAL_TOKENS_URL=redis:6379/6
POLYAXON_REDIS_TTL_URL=redis:6379/7
POLYAXON_REDIS_HEARTBEAT_URL=redis:6379/8
POLYAXON_REDIS_GROUP_CHECKS_URL=redis:6379/9
POLYAXON_REDIS_STATUSES_URL=redis:6379/10
POLYAXON_REDIS_TTL_URL=redis:6379/7
POLYAXON_HEARTBEAT_URL=redis:6379/8
```

编写 docker-compose 文件，用于启动 Polyaxon：

```
version: '3.4'

x-defaults: &defaults
  restart: unless-stopped
  networks:
    - polyaxon-compose
  depends_on:
    - redis
    - postgres
  env_file:
```

```yaml
    - base.env
    - components.env
    - .env

services:
  postgres:
    restart: unless-stopped
    image: postgres:9.6-alpine
    environment:
      POSTGRES_USER: "polyaxon"
      POSTGRES_PASSWORD: "polyaxon"
    volumes:
      - polyaxon-postgres:/var/lib/postgresql/data
    networks:
      - polyaxon-compose

  redis:
    image: redis:5.0.5-alpine
    networks:
      - polyaxon-compose

  web:
    <<: *defaults
    image: polyaxon/polyaxon-api:latest
    command: ["--disable-plugins"]
    ports:
      - "8000:80"
      - "8001:443"

  worker:
    <<: *defaults
    image: polyaxon/polyaxon-worker:0.5.1

  beat:
    <<: *defaults
    image: polyaxon/polyaxon-beat:0.5.1

  sync-db:
    <<: *defaults
```

```yaml
    restart: "no"
    image: polyaxon/polyaxon-manage:0.5.1
    command: ["migrate"]
    environment:
      POLYAXON_DB_NO_CHECK: 1

  create-user:
    <<: *defaults
    restart: "no"
    image: polyaxon/polyaxon-manage:0.5.1
    command: ["createuser --username=root --email=toor@loca.com --password=root --superuser --force"]

volumes:
  polyaxon-postgres:
    external: true

networks:
  polyaxon-compose:
```

配置完成后可以通过 docker-compose up -d 命令启动 Polyaxon。Polyaxon 的模型训练界面如图 7-2 所示。

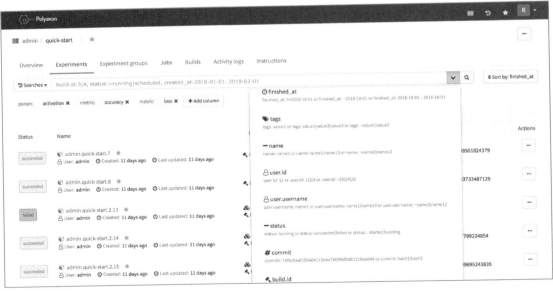

图 7-2

Polyaxon 训练任务的可视化界面如图 7-3 所示。

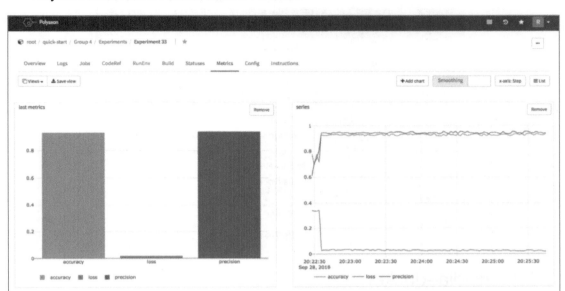

图 7-3

7.2 搭建 AI 平台的技术点

除了使用 Polyaxon 和 Kubeflow 这些成型的 AI 平台，我们也可以基于 Kubernetes 搭建属于自己的 AI 平台。不过搭建 AI 平台所使用的技术非常多，笔者在工作中设计了一套基于 Kubernetes 搭建的 AI 平台——AlphaMind，AlphaMind 在搭建的过程中使用了非常多的组件，也遇到了非常多的问题。本节分享搭建 AlphaMind 过程中的一些技术点，AlphaMind 平台的技术栈如图 7-4 所示。

在这套方案中，AI 平台被拆分为 AI 平台底座与 AI 平台，AI 平台底座负责连接 AI 平台与底层的系统基石，为 AI 平台业务屏蔽底层的复杂性。系统借助 Kubernetes 进行资源调度及管理，所有的模型训练任务和推理任务都以容器的形态出现。借助 Kubernetes 的弹性伸缩和资源纳管能力，我们能够非常方便地搭建一套生产级的 AI 平台。

下面介绍 AI 平台搭建过程中的常用技术点。

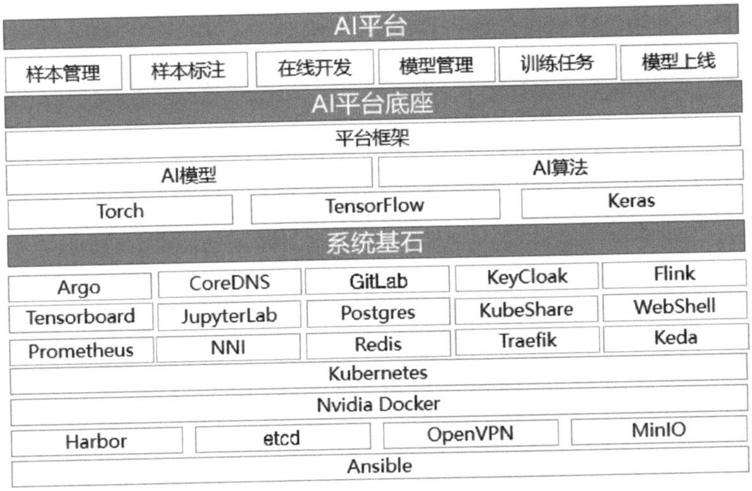

```
                    AI平台
  ┌──────┬──────┬──────┬──────┬──────┬──────┐
  │样本管理│样本标注│在线开发│模型管理│训练任务│模型上线│
  └──────┴──────┴──────┴──────┴──────┴──────┘
                  AI平台底座
  ┌─────────────────────────────────────────┐
  │                平台框架                  │
  ├─────────────────────┬───────────────────┤
  │       AI模型         │       AI算法      │
  ├──────────┬──────────┴──────┬────────────┤
  │  Torch   │    TensorFlow   │   Keras    │
  └──────────┴─────────────────┴────────────┘
                   系统基石
  ┌────────┬────────┬────────┬────────┬────────┐
  │  Argo  │CoreDNS │ GitLab │KeyCloak│ Flink  │
  ├────────┼────────┼────────┼────────┼────────┤
  │Tensorboard│JupyterLab│Postgres│KubeShare│WebShell│
  ├────────┼────────┼────────┼────────┼────────┤
  │Prometheus│  NNI  │ Redis  │Traefik │  Keda  │
  ├────────┴────────┴────────┴────────┴────────┤
  │              Kubernetes                    │
  ├────────────────────────────────────────────┤
  │             Nvidia Docker                  │
  ├──────────┬────────┬──────────┬─────────────┤
  │  Harbor  │  etcd  │ OpenVPN  │   MinIO     │
  ├──────────┴────────┴──────────┴─────────────┤
  │                Ansible                     │
  └────────────────────────────────────────────┘
```

图 7-4

7.2.1 nvidia-docker

1. 安装 nvidia-docker

AI 模型的训练和推理经常会用到 GPU 资源。Docker 在默认的情况下无法使用 GPU 资源，为了让 Docker 能够使用 GPU 资源，需要为 Docker 安装 nvidia-docker。

在操作系统中安装 Nvidia 驱动后，执行以下命令可以完成 nvidia-docker 的安装：

```
yum clean expire-cache
yum install -y nvidia-docker2
```

nvidia-docker 安装完毕后，在/etc/docker/daemon.json 中添加如下内容并重启 Docker：

```
"default-runtime": "nvidia",
"runtimes": {
  "nvidia": {
    "path": "/usr/bin/nvidia-container-runtime",
    "runtimeArgs": []
  }
}
```

使用 tensorflow-gpu 镜像查看 nvidia-docker 是否安装成功：

```
docker run --rm -it tensorflow/tensorflow:1.15.4-gpu-py3 bash
```

进入容器后执行 nvidia-smi 命令，可以看到在容器内已经能够调用 GPU 资源：

```
root@d4a72f36ba6f:/# nvidia-smi
Mon Apr  5 02:36:34 2021
+-----------------------------------------------------------------------------+
| NVIDIA-SMI 410.48                 Driver Version: 440.59                    |
|-------------------------------+----------------------+----------------------+
| GPU  Name        Persistence-M| Bus-Id        Disp.A | Volatile Uncorr. ECC |
| Fan  Temp  Perf  Pwr:Usage/Cap|         Memory-Usage | GPU-Util  Compute M. |
|===============================+======================+======================|
|   0  GeForce RTX 208...  Off  | 00000000:0B:00.0 Off |                  N/A |
| 18%   38C    P8    13W / 250W |      0MiB / 11019MiB |      0%      Default |
+-------------------------------+----------------------+----------------------+
|   1  GeForce RTX 208...  Off  | 00000000:13:00.0 Off |                  N/A |
| 18%   44C    P8     1W / 250W |      0MiB / 11019MiB |      0%      Default |
+-------------------------------+----------------------+----------------------+
```

若要验证是否能够使用 cuda，则可以进入 Python 交互式界面，输入以下命令：

```
import tensorflow as tf
tf.test.is_gpu_available()
```

2. nvidia-docker 的调度流程与系统兼容性

nvidia-docker 由以下组件组成：

- nvidia-docker2；
- nvidia-container-runtime；
- nvidia-container-toolkit；
- libnvidia-container。

通过 nvidia-docker 调度 GPU 资源的流程如图 7-5 所示。

值得注意的是，并不是所有的 Linux 发行版都支持 nvidia-docker，所以在选择操作系统的时候，需要关注是否在 ARM 架构或者 ppc64le 架构上执行 nvidia-docker。表 7-1 为常用的 Linux 发行版与 nvidia-docker 兼容性的对照表。

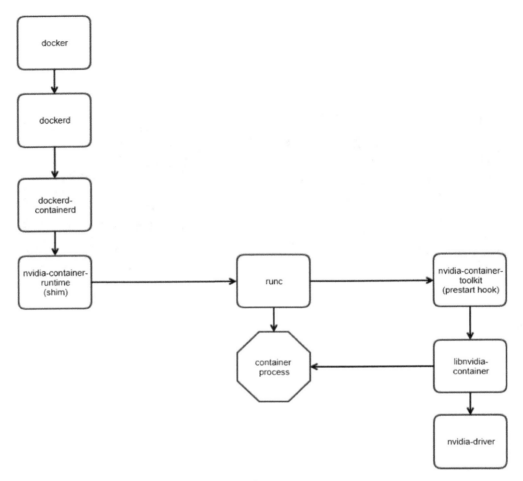

图 7-5

表 7-1

操作系统	版本号	AMD64/x86_64	ppc64le	ARM64/AArch64
Open Suse Leap 15.0	sles15.0	√		
Open Suse Leap 15.1	sles15.0	√		
Debian Linux 9	debian9	√		
Debian Linux 10	debian10	√		
Centos 7	centos7	√	√	
Centos 8	centos8	√	√	√
RHEL 7.4	rhel7.4	√	√	
RHEL 7.5	rhel7.5	√	√	

续表

操作系统	版本号	AMD64/x86_64	ppc64le	ARM64/AArch64
RHEL 7.6	rhel7.6	√	√	
RHEL 7.7	rhel7.7	√	√	
RHEL 8.0	rhel8.0	√	√	√
Ubuntu 16.04	ubuntu16.04	√	√	
Ubuntu 18.04	ubuntu18.04	√	√	√
Ubuntu 20.04	ubuntu20.04	√	√	√

可以看到，在操作系统的选择上，CentOS 8、RHEL 8.0 和 Ubuntu 20.04 是不错的选择，nvidia-docker 对这两个版本的操作系统的兼容性都非常好。不过值得注意的是，随着 CentOS 的版本发布规则出现了变化（以 CentOS Stream 模式发布），对系统稳定性有要求的读者，建议采用 RHEL 8.0 或者 Ubuntu 20.04。

7.2.2　nvidia-device-plugin

除了 Docker 在默认情况下无法使用 GPU 资源，Kubernetes 在默认情况下也是无法使用 GPU 资源的，为 Kubernetes 安装 nvidia-device-plugin 之后才能够管理 GPU 资源。

编写 nvidia-device-plugin.yml 文件，加入以下内容：

```yaml
apiVersion: apps/v1
kind: DaemonSet
metadata:
  name: nvidia-device-plugin-daemonset
  namespace: kube-system
spec:
  selector:
    matchLabels:
      name: nvidia-device-plugin-ds
  updateStrategy:
    type: RollingUpdate
  template:
    metadata:
      annotations:
        scheduler.alpha.kubernetes.io/critical-pod: ""
      labels:
        name: nvidia-device-plugin-ds
```

```yaml
spec:
  tolerations:
  - key: CriticalAddonsOnly
    operator: Exists
  - key: nvidia.com/gpu
    operator: Exists
    effect: NoSchedule
  priorityClassName: "system-node-critical"
  containers:
  - image: nvidia/k8s-device-plugin:1.0.0-beta6
    name: nvidia-device-plugin-ctr
    securityContext:
      allowPrivilegeEscalation: false
      capabilities:
        drop: ["ALL"]
    volumeMounts:
      - name: device-plugin
        mountPath: /var/lib/kubelet/device-plugins
  volumes:
    - name: device-plugin
      hostPath:
        path: /var/lib/kubelet/device-plugins
```

然后执行 kubectl apply -f ./ nvidia-device-plugin.yml 进行安装，安装完成后查看 GPU 是否成功被 Kubernetes 纳管：

```
kubectl describe nodes|grep -A5 -B5 gpu
```

```
  cpu:                40
  ephemeral-storage:  51175Mi
  hugepages-1Gi:      0
  hugepages-2Mi:      0
  memory:             51352292Ki
  nvidia.com/gpu:     2
  pods:               110
Allocatable:
  cpu:                40
  ephemeral-storage:  50977832921
  hugepages-1Gi:      0
  hugepages-2Mi:      0
  memory:             51352292Ki
  nvidia.com/gpu:     2
  pods:               110
```

可以看到 Kubernetes 已经能够对 GPU 资源进行管理了，至此，我们已经能够在 Kubernetes 上使用 GPU 资源了。

7.2.3　KubeShare——显卡资源调度

由于显卡资源相对昂贵，如何充分地利用显卡资源是一个非常重要的技术点。默认情况下，Kubernetes 对显卡资源的限制有两种，一种是一个 Pod 独占显卡，另一种是非独占。一个 Pod 能看到所有的显卡，这种隔离方式对显卡资源的限制在生产环境下往往是不够的。基于 Kubernetes 的显卡共享策略有很多，KubeShare 是其中一种应用侵入性较小的显卡共享解决方案。

安装 KubeShare：

```
kubectl create -f ./KubeShare/v0.9/crd.yaml
kubectl create -f ./KubeShare/v0.9/device-manager.yaml
kubectl create -f ./KubeShare/v0.9/scheduler.yaml
```

为容器开启显存共享的特性：

```
apiVersion: kubeshare.nthu/v1
kind: SharePod
metadata:
  name: sharepod
  annotations:
    "kubeshare/gpu_request": "0.5"
    "kubeshare/gpu_limit": "1.0"
    "kubeshare/gpu_mem": "1073741824"
spec:
  containers:
  - name: cuda
    image: nvidia/cuda:9.0-base
    command: ["nvidia-smi", "-L"]
    resources:
      limits:
        cpu: "1"
        memory: "500Mi"
```

在此配置下，sharepod 声明了需要使用 0.5 个 GPU，最多使用 1 个 GPU，其中 GPU 的最大使用内存为 1GB。可以看到，KubeShare 是使用注解的方式配置的，对原有的应用配置的改动非常小。

GPU 共享的场景在训练任务的时候用得较少，因为在训练环节我们期望任务执行得越快越好。GPU 共享的能力在 AI 模型开发环节使用得较多，当 AI 开发人员的数量比显卡数量多的时候，使用显卡共享的模式能够大幅提高显卡资源的利用率，这也是 AI 平台非常重要的一个能力。

7.2.4　AI 算法插件框架设计

使用 Polyaxon 和 Kubeflow 的时候，都需要用到它们的 SDK，但是这样容易给算法开发人员带来一些不便。对于算法开发人员来说，假如不需要关注 AI 平台的 SDK，那么按照一定的约定编写代码，这份代码就能被 AI 平台自行调度，这样可以缩短不少 AI 模型开发的时间。所以在设计 AI 平台的时候，算法插件框架的设计也显得非常重要。下面提供一种 AI 算法插件的设计方式：

```python
import os

class SeriesDetection():
    def __init__(self, job_context):
        self.job_context = job_context
        self.asserts = {}

    def pre_train(self):
        pass

    def train(self):
        pass

    def evaluate(self):
        return 1, 1

    def init_model(self):
        pass

    def predict(self, data):
        pass
```

首先，我们把训练和推理两个生命周期分开，约定训练的生命周期为__init__→pre_train→

train→evaluate。在 pre_train 函数中，主要是做一些数据预处理的工作，在 train 环节，则开启训练模式，并将训练好的模型存放至指定的位置，如果框架在指定位置上发现了模型，则将模型持久化到分布式文件系统中。最后的 evaluate 函数用于自我评估模型的精度，返回的值会被 AI 平台的 NNI 工具包作为算法自动调优的依据。而推理的生命周期为 __init__→init_model→predict，init_model 全局被执行一次，用于初始化 AI 模型，predict 函数实现 AI 模型的推理，每次推理都会被调用。一个基本的 AI 算法框架的插件就约定好了。

AI 平台框架调度流程如图 7-6 所示。

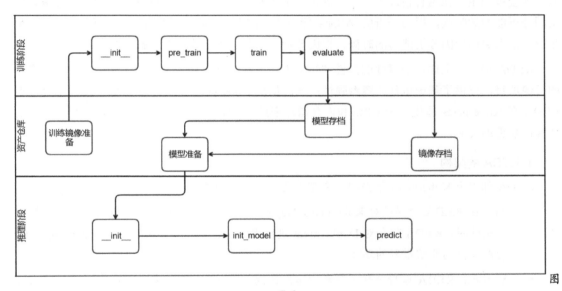

图 7-6

AI 插件的函数约定好了，这些插件是怎么运作的呢？一个插件变成可用的 AI 能力会经过如下几个环节，首先是训练阶段：

（1）插件包和框架被打包成一个容器，交给 Kubernetes 以 Job 的方式运行。

（2）启动的 Job 会自动下载训练数据集，并开始模型训练。

（3）训练好的模型存放到指定位置，框架会自动将模型存放至分布式系统中。

（4）返回评估结果，供后续自动化调参时使用。

然后是推理阶段：

（1）启动推理服务，将插件包和框架打包好的容器以 Kubernetes 的 Deployment 形式启动。

（2）自动调用插件包的 init_model 函数，初始化 AI 模型。

（3）提供 RESTful/RPC 服务，当收到参数的时候，调用 predict 函数。

7.2.5 KEDA——基于事件的弹性伸缩框架

每个 AI 能力上线后，无论是否被其他系统调用，都会占用内存或显存等资源，但并不是所有的模型都需要长期开启，随时提供服务。如何最大限度地提高 AI 平台的整体资源利用率，对外提供更好的 AIOps 推理服务，是 AI 平台设计者所需要关注的重要问题。

以根因分析为例，只有当告警发生或需要对历史数据做根因分析的时候，根因分析的推理服务才会被调用。在这种情况下，平台就需要提供基于事件的弹性伸缩能力，当没有服务调用根因分析能力的时候，根因分析的 AIOps 能力在集群中没有启动任何服务，当有请求进入 AI 平台，请求调用根因分析能力的时候，根因分析的服务自动上线，对外提供服务。

KEDA 提供了上述基于事件的弹性伸缩能力，它是一个由 Red Hat 和 Microsoft 的团队合作的开源项目，提供了基于事件的容器弹性扩缩。虽然 Kubernetes 也提供了弹性伸缩的功能，但 KEDA 在 Kubernetes 原生基于 CPU 和内存指标来扩/缩容器的模式下进行了扩展，使得容器的弹性伸缩更加灵活。

1. KEDA 的架构

KEDA 部署于 Kubernetes 集群中，它充当两个主要的角色：

- Agent：KEDA 会提供对 Kubernetes Deployment 的基于事件的缩放能力，当满足特定的条件时，KEDA 就会对 Deployment 进行操作，这是安装完 KEDA 后，keda-operator 容器提供的非常重要的功能。
- Metrics：KEDA 自身公开了丰富的事件数据，如队列长度、Kafka 的 lag 之类的事件，Keda 基于 Metrics 所监控的指标数据提供了一些"开箱即用"的弹性伸缩能力。

KEDA 的架构如图 7-7 所示。

从架构图中，可以看出 KEDA 的工作流程，KEDA 部署到 Kubernetes 上之后，会观察我们所配置的触发器，当满足条件之后，调用 Kubernetes 的能力，对容器进行弹性的伸缩，这也就是为什么 KEDA 能拥有"scale to zero"（在无应用的时候，让应用的实例为零）的能力了。

2. KEDA 的部署

有多种方式可以部署 KEDA，下面介绍 Deployment 和 Helm 两种部署模式。

使用 Deployment 部署 KEDA 是一种简单便捷的方式，KEDA 所提供的 Deployment 中已经包含所有需要部署的资源，直接启用即可：

```
kubectl apply -f ./keda-2.1.0.yaml
```

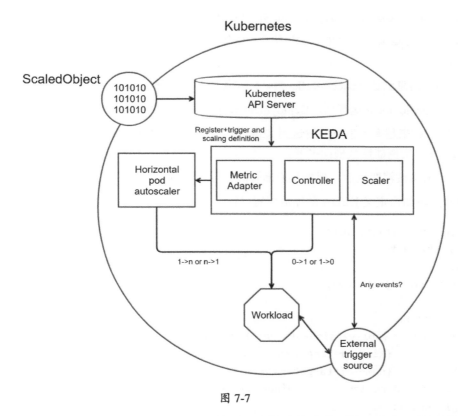

图 7-7

使用 Deployment 的方式部署 KEDA 比较适用于快速实验的环境，生产环境下需要对配置进行个性化的调整，使用 Helm 部署是一种更优的方式：

```
kubectl create namespace keda
helm install keda kedacore/keda --namespace keda
```

部署完成后，在命令行中输入 kubectl get po -n keda，看到如下输出则表示 KEDA 已经正常运行了：

```
NAME                                       READY   STATUS    RESTARTS   AGE
keda-metrics-apiserver-77566dc65-plq9c     1/1     Running   0          20h
keda-operator-749f57d796-t7fxb             1/1     Running   2          20h
```

3. KEDA 的基本使用方法

对 Deployment 和 StatefulSet 进行基于事件的弹性伸缩是 KEDA 常见的一个应用场景。KEDA 可以对事件源进行监控，并对 Deployment 或 StatefulSet 进行弹性伸缩。

例如，如果将 Apache Kafka 的 Topic 作为 KEDA 的事件源，则 KEDA 的工作流程如下：

（1）当 Apache Kafka 的 Topic 没有收到任何消息的时候，KEDA 将目标的 Deployment 实例调整为 0。

（2）当有消息到达 Apache Kafka 后，KEDA 会检测到事件，并部署 Deployment。

（3）当 Deployment 运行后，Deployment 开始从 Kafka 中消费数据。

（4）随着越来越多的消息被发送到 Kafka 中，KEDA 根据我们所配置的条件将扩展信息推送给 HPA，使得实例能够被横向扩展。

（5）当消息被消费完后，根据我们部署的条件，Deployment 的数量会逐渐缩小至 0。

下面看一下如何定义 KEDA 的缩放配置模板：

```
apiVersion: keda.sh/v1alpha1
kind: ScaledObject
metadata:
  name: {scaled-object-name}
spec:
  scaleTargetRef:
    apiVersion:    {api-version-of-target-resource}
    kind:          {kind-of-target-resource}
    name:          {name-of-target-resource}
    envSourceContainerName: {container-name}
  pollingInterval: 30
  cooldownPeriod:  300
  minReplicaCount: 0
  maxReplicaCount: 100
  advanced:
    restoreToOriginalReplicaCount: true/false
    horizontalPodAutoscalerConfig:
      behavior:
        scaleDown:
          stabilizationWindowSeconds: 300
          policies:
          - type: Percent
            value: 100
            periodSeconds: 15
  triggers:
    ......
```

scaleTargetRef 配置区定义了需要弹性伸缩的资源，主要的配置为 name，声明了我们需要弹性伸缩的资源是哪个。需要注意的是，所指向的 Deployment 必须和 SclaeObject 在同一个命名空间。

接下来需要关注配置项中的如下参数：

- pollingInterval：定义了 KEDA 检测触发弹性伸缩事件源的周期，默认是 30 秒检查一次，需要根据弹性伸缩的要求进行合理的配置；
- cooldownPeriod：定义了事件源结束（如 Kafka 队列中没有消息）后将资源缩放为 0 的时间，默认是 300 秒；
- minReplicaCount：此配置项默认为 0，代表当事件结束后，KEDA 需要将 Deployment 缩放至多少；
- maxReplicaCount：默认为 100，代表 KEDA 最大能将资源缩放至多少；
- restoreToOriginalReplicaCount：默认为 false，声明了当 Deployment 被删除后，是否将 Deployment 缩放为被删除前的副本数。

下面给出了将 KEDA 应用于 Kafka 消费者的一种配置方式，目标 Deployment 在 Kafka Consumer 没有从 Kafka 中获取任何消息的时候，不启动任何 Pod，当 Kafka 中有消息的时候，开始对容器进行动态缩放：

```
apiVersion: keda.k8s.io/v1alpha1
kind: ScaledObject
metadata:
  name: kafka-scaledobject
  namespace: keda
  labels:
    deploymentName: kafka-consumer-deployment
spec:
  scaleTargetRef:
    deploymentName: kafka-consumer-deployment
  pollingInterval: 5
  minReplicaCount: 1
  maxReplicaCount: 10
  triggers:
  - type: kafka
    metadata:
      brokerList: 192.168.1.4:9092
      consumerGroup: order-shipper
```

```
    topic: test
    lagThreshold: "10"
    consumer group
```

4. 在 KEDA 中调度长周期任务

对于大多数场景，KEDA 可以轻松自如地对应用进行弹性伸缩，但是有一种特殊的情况需要引起注意，这种情况就是长周期任务。假设我们部署了一套基于 Kafka 消息队列中的消息进行弹性伸缩的服务，由于任务的特殊性，每个消息需要花费 3 个小时才能完成处理，当越来越多的消息到达 Kafka 时，KEDA 会对 Kafka 消费者的数量进行动态调整，将副本数量扩展为 8 个。一段时间后，HPA 计划将 8 个副本缩减为 4 个。这时问题来了，HPA 无法准确地判断应该缩小哪一个副本以降低 Pod 的副本规模，假如随机对副本进行缩小，则很可能导致一个已经处理了 2 个小时的副本被删除。

有两种方式可以解决上述问题：第一种方式是利用容器的生命周期的特点。Kubernetes 提供了一些生命周期的 Hook 可以用来延迟 Pod 被立刻终止。在 Kubernetes 让 Pod 被终止的时候，会发送一个 SIGTERM 的信号，但 Deployment 可以选择延迟终止，直到当前的任务处理完成后，再终止容器。另外一个可选的方法是使用 Kubernetes Jobs 运行任务，而不是 Deployment。

以下为使用 KEDA 调度长周期任务的配置示例：

```
apiVersion: keda.sh/v1alpha1
kind: ScaledJob
metadata:
  name: {scaled-job-name}
spec:
  jobTargetRef:
    parallelism: 1
    completions: 1
    activeDeadlineSeconds: 600
    backoffLimit: 6
    template:
  pollingInterval: 30
  successfulJobsHistoryLimit: 5
  failedJobsHistoryLimit: 5
  envSourceContainerName: {container-name}
  maxReplicaCount: 100
  scalingStrategy:
    strategy: "custom"
```

```yaml
    customScalingQueueLengthDeduction: 1
    customScalingRunningJobPercentage: "0.5"
  triggers:
  ......
```

ScaledJob 与 ScaledObject 最大的不同是缩放任务从 Deployment 转变成了 Job，不再需要关联一个具体的 Deployment，直接在 ScaledJob 中声明任务即可。下面给出一个示例，采用 ScaleJob 的模式弹性调度任务：

```yaml
apiVersion: keda.sh/v1alpha1
kind: ScaledJob
metadata:
  name: rabbitmq-consumer
  namespace: default
spec:
  jobTargetRef:
    template:
      spec:
        containers:
        - name: rabbitmq-client
          image: tsuyoshiushio/rabbitmq-client:dev3
          imagePullPolicy: Always
          command: ["receive", "amqp://user:PASSWORD@rabbitmq.default.svc.cluster.local:5672", "job"]
          envFrom:
            - secretRef:
                name: rabbitmq-consumer
        restartPolicy: Never
    backoffLimit: 4
  pollingInterval: 10
  maxReplicaCount: 30
  successfulJobsHistoryLimit: 3
  failedJobsHistoryLimit: 2
  scalingStrategy:
    strategy: "custom"
    customScalingQueueLengthDeduction: 1
    customScalingRunningJobPercentage: "0.5"
  triggers:
```

```
- type: rabbitmq
  metadata:
    queueName: hello
    host: RabbitMqHost
    queueLength : '5'
```

5. KEDA 的弹性伸缩触发器

KEDA 提供了非常多的弹性伸缩触发器，当应用达到某个状态时就触发自动伸缩的操作。下面看一下 KEDA 中常用的一些触发器。

1）Kafka 触发器

AI 平台中会有非常多的数据流处理任务，例如，将监控指标的数据从 Kafka 流入平台，训练时序预测或异常检测等 AIOps 能力。因此，基于 Kafka 的弹性伸缩触发器也是常用的一种触发器。以下代码展示了使用 Kafka 触发器的方法：

```
apiVersion: keda.sh/v1alpha1
kind: ScaledObject
metadata:
  name: kafka-scaledobject
  namespace: default
spec:
  scaleTargetRef:
    name: azure-functions-deployment
  pollingInterval: 30
  triggers:
  - type: kafka
    metadata:
      bootstrapServers: localhost:9092
      consumerGroup: my-group
      topic: test-topic
      lagThreshold: "50"
      offsetResetPolicy: latest
```

Kafka 触发器中的关键配置项如下：

- bootstrapServers：Kafka 的地址；
- consumerGroup：消费者的分组；

- topic：监控的 Topic，当有消息到达 Topic 时，触发器会触发弹性伸缩的能力；
- lagThreshold：触发缩放行为的目标值；
- offsetResetPolicy：消费者的偏移策略，默认为 latest。

2）Prometheus 触发器

当 AI 平台中的监控数据采用 Prometheus 进行存放的时候，我们可以使用 Prometheus 触发器进行 AI 推理服务的弹性缩放，将推理服务的关键 KPI 如模型推理耗时、推理请求数量等写入 Prometheus，使用 KEDA 的 Prometheus 触发器进行推理服务的弹性缩放。

使用 Prometheus 触发器的示例如下：

```
apiVersion: keda.sh/v1alpha1
kind: ScaledObject
metadata:
  name: prometheus-scaledobject
  namespace: default
spec:
  scaleTargetRef:
    name: my-deployment
  triggers:
  - type: prometheus
    metadata:
      serverAddress: http://prometheus-svc:9090
      metricName: http_requests_total
      threshold: '100'
      query: sum(rate(http_requests_total{deployment="my-deployment"}[2m]))
```

Prometheus 触发器中的关键配置项如下：

- serverAddress：Prometheus 的地址；
- metricName：目标指标的名称；
- threshold：触发弹性伸缩的阈值；
- query：指标查询语句。

7.2.6　Argo Workflow——云原生的工作流引擎

在 AI 模型的训练过程中，涉及多个 AI 模型串联训练的场景。以日志异常检测算法为例，涉及将日志模板提取、LSTM 算法进行串联等场景，这时就需要用到工作流引擎，将两个训练任务进行串联。云原生工作流 Argo Workflow 是一个不错的选择，它的特点如下：

- 使用 Kubernetes 自定义资源定义工作流，工作流中的每个步骤都是一个容器；
- 将多步骤工作流建模为一系列任务，或者使用有向无环图（DAG）描述任务之间的依赖关系；
- 可以在短时间内轻松运行用于机器学习或数据处理的计算密集型作业。

1. 部署 Argo Workflow

在开始讲解 Argo Workflow 的部署和使用方式之前，先对 Argo Workflow 的关键概念进行讲解，便于我们更加容易地掌握 Argo Workflow 的使用方式，如表 7-2 所示。

表 7-2

关键概念	描述
Workflow	Kubernetes 的一种资源，用于定义一个或多个执行流水线
Template	包含对 step、steps 或 dag 的描述
Step	Workflow 中的一个基础步骤，一般包括输入和输出两个环节
Steps	包含多个 Step
EntryPoint	执行 Workflow 的入口，第一个需要被执行的任务
Cluster Workflow Template	包含集群中可以被重复使用 Workflow
Inputs	Step 所接收的输入参数
Outputs	Step 所接收的输出参数
Parameters	Workflow 所接收的参数，可以是对象、字符串、数组、布尔值
Artifacts	容器保存的文件
Artifact Repository	文件仓库
Executor	启动容器的方式，例如 Docker、PNS

了解了上面的关键概念之后，接下来就可以部署 Argo Workflow 了。我们直接使用 Deployment 的方式部署 Argo Workflow。

```
kubectl create ns argo
kubectl apply -n argo -f ./quick-start-postgres.yaml
```

接下来访问 127.0.0.1:2746，当看到如图 7-8 所示的界面时，表示 Argo Workflow 已经部署完成了。

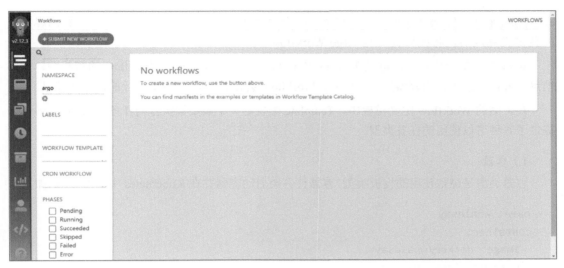

图 7-8

2. Argo Workflow 快速入门

下面通过一些示例及 Argo Workflow 的资源定义模板学习如何使用 Argo Workflow。

Workflow 是 Argo Workflow 中最重要的资源，它提供了两个功能：

- 定义了 Workflow 的执行流程；
- 存储了 Workflow 的状态信息。

下面是一个简单的 Workflow 示例文件：

```
apiVersion: argoproj.io/v1alpha1
kind: Workflow
metadata:
  generateName: hello-world
spec:
  entrypoint: whalesay
  templates:
  - name: whalesay
    container:
      image: docker/whalesay
      command: [cowsay]
      args: ["hello world"]
```

首先定义了使用的资源类型为 Workflow，在 Workflow 资源类型配置中的 spec 部分，使用

entrypoint 定义了 Workflow 的第一个被执行的步骤。最后在 templates 中定义了需要执行的容器。配置定义完成后，将配置提交至 Argo Workflow。

Workflow 提交成功后，可以在 Argo Workflow 的详情界面看到 Workflow 的拓扑图、概要信息、容器日志等，当容器执行完成后，Workflow 会在 Output 中输出执行的日志详细记录。

在上面的 Workflow 定义模板中，Template 配置定义了需要被执行的任务，Argo Workflow 提供了 6 种可以使用的任务类型。

1）容器

容器类型是经常使用的模板类型，容器任务会启动容器并在 Kubernetes 中调度。示例如下：

```yaml
- name: whalesay
  container:
    image: docker/whalesay
    command: [cowsay]
    args: ["hello world"]
```

2）脚本

脚本任务和容器任务的区别在于增加了"source: 配置属性"，允许我们将需要执行的脚本定义在配置文件中，无须每次都重新编译镜像，脚本的执行结果也会被自动存放至 Argo Workflow 的变量中。示例如下：

```yaml
- name: gen-random-int
  script:
    image: python:alpine3.6
    command: [python]
    source: |
      import random
      i = random.randint(1, 100)
      print(i)
```

3）资源

资源类型的任务用于对集群中的资源进行操作，可用于获取、创建、应用、删除、替换或更新集群上的资源。示例如下：

```yaml
- name: k8s-owner-reference
  resource:
    action: create
```

```
manifest: |
  apiVersion: v1
  kind: ConfigMap
  metadata:
    generateName: owned-eg-
  data:
    some: value
```

4）暂停

暂停任务可以对 Workflow 的执行流程进行阻塞，让 Workflow 停止指定的时间，或者要求直到用户手动操作后，Workflow 才恢复执行。示例如下：

```
- name: delay
  suspend:
    duration: "20s"
```

5）Step

Step 类型的任务可以帮助我们定义 Workflow 的执行顺序。例如，在下面的示例中，step1 会先被执行，执行完成后，step2a 和 step2b 会被并行执行。也可以配置 when 属性让任务有条件地执行，如 step2a 执行完成后再执行 step2b。

```
- name: hello-hello-hello
  steps:
  - - name: step1
      template: prepare-data
  - - name: step2a
      template: run-data-first-half
    - name: step2b
      template: run-data-second-half
```

6）DAG

DAG 类型的任务能够帮助我们将任务定义为一张依赖关系图，可以设置在开始特定任务之前必须完成的其他任务，那些没有任何依赖关系的任务将被立即运行。例如，在下面的示例中，A 任务会被先执行，由于 B 和 C 任务都配置了 dependencies 属性，要求 A 执行完成后才执行 B 和 C，所以 A 任务执行完成后，B 和 C 任务都会被执行，D 任务依赖于 B 和 C 任务，所以 D 任务会在 B 和 C 任务都执行完成后执行。

```
- name: diamond
```

```
dag:
  tasks:
  - name: A
    template: echo
  - name: B
    dependencies: [A]
    template: echo
  - name: C
    dependencies: [A]
    template: echo
  - name: D
    dependencies: [B, C]
    template: echo
```

以下示例展示了 DAG 任务的具体配置方式：

```
apiVersion: argoproj.io/v1alpha1
kind: Workflow
metadata:
  generateName: dag
spec:
  entrypoint: startup
  templates:
  - name: echo
    inputs:
      parameters:
      - name: message
    container:
      image: alpine:3.7
      command: [echo, "{{inputs.parameters.message}}"]
  - name: startup
    dag:
      tasks:
      - name: A
        template: echo
        arguments:
          parameters: [{name: message, value: A}]
      - name: B
        dependencies: [A]
        template: echo
        arguments:
```

```
          parameters: [{name: message, value: B}]
    - name: C
      dependencies: [A]
      template: echo
      arguments:
          parameters: [{name: message, value: C}]
    - name: D
      dependencies: [B, C]
      template: echo
      arguments:
          parameters: [{name: message, value: D}]
```

将上述示例提交给 Argo Workflow 后，会生成如图 7-9 所示的拓扑图。

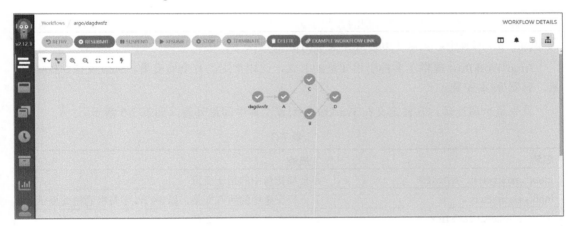

图 7-9

我们可以更加清晰地看到，通过使用 Step 任务，可以帮助我们定义更加复杂的工作流执行方式。

了解如何配置 Argo Workflow 的工作流之后，我们还需要了解如何使用 Argo Workflow 中的变量，先看如下示例：

```
apiVersion: argoproj.io/v1alpha1
kind: Workflow
metadata:
  generateName: hello-world-parameters-
spec:
  entrypoint: whalesay
  arguments:
```

```
parameters:
  - name: message
    value: hello world
templates:
  - name: whalesay
    inputs:
      parameters:
        - name: message
    container:
      image: docker/whalesay
      command: [ cowsay ]
      args: [ "{{inputs.parameters.message}}" ]
```

上述示例中多了一个比较特别的参数 args，并且声明了这是一个输入参数，名称为 message。在 arguments 配置中，定义了 message 变量的值为 hello world。

Argo Workflow 提供了多种引用变量的方式，比较常用的有全局变量、Step 变量、DAG 变量、容器/脚本变量。

首先是全局变量，它被定义在 Workflow 配置文件中的最外层，如表 7-3 所示。

表 7-3

名称	描述
inputs.parameters.<NAME>	全局变量中的指定变量
inputs.parameters	全局变量中的所有变量，以 JSON 字符串的模式提供
inputs.artifacts.<NAME>	全局变量中的指定文件

Step 的可用变量比全局变量多得多，此处仅挑选比较常用的变量进行介绍，如表 7-4 所示。

表 7-4

名称	描述
steps.<STEPNAME>.id	Step 的唯一 ID
steps.<STEPNAME>.status	Step 的状态
steps.<STEPNAME>.exitCode	Step 的运行结束状态，用于判断是否正常运行
steps.<STEPNAME>.outputs.result	Step 执行完成后的输出结果
steps.<STEPNAME>.outputs.parameters.<NAME>	当上一个 Step 使用了 withItems 或 withParams 的配置后，当前 Step 以 Map 的模式使用上一个 Step 的输出参数
steps.<STEPNAME>.outputs.artifacts.<NAME>	上一个 Step 输出的文件

DAG 的常用变量如表 7-5 所示。

表 7-5

名称	描述
tasks.<TASKNAME>.id	DAG 任务的唯一 ID
tasks.<TASKNAME>.status	DAG 任务的状态
tasks.<TASKNAME>.exitCode	DAG 任务的运行结束状态，用于判断是否正常运行
tasks.<TASKNAME>.outputs.result	DAG 任务执行完成后的输出结果
tasks.<TASKNAME>.outputs.parameters.<NAME>	当上一个任务使用 withItems 或 withParams 的配置后，当前任务以 Map 的模式使用上一个任务的输出参数
tasks.<TASKNAME>.outputs.artifacts.<NAME>	上一个任务输出的文件

常用的变量、容器/脚本变量如表 7-6 所示。

表 7-6

名称	描述
pod.name	容器名称
retries	任务重试次数
inputs.artifacts.<NAME>.path	输入的文件变量的路径
outputs.artifacts.<NAME>.path	输出的文件变量的路径
outputs.parameters.<NAME>.path	输出变量的路径

3. Argo Workflow 典型配置

1）在任务之间读取文件

AI 模型训练的第一步通常是数据处理，数据处理完成后，会生成一些预处理后的文件，然后交给训练任务进行训练。训练任务结束后，会产生模型文件。最后一步是将上一步生成的模型文件进行归档。所以在任务之间读取文件是 AI 模型训练中的一个非常典型且常见的场景。

下面的配置展示了如何在多个任务之间读取文件。其中关键的配置是在 Step 配置中定义了需要使用 print-message 变量，Argo Workflow 会根据使用者的要求将变量传递给下一个任务。

```
apiVersion: argoproj.io/v1alpha1
kind: Workflow
metadata:
```

```yaml
  generateName: artifact-passing-
spec:
  entrypoint: artifact-example
  templates:
  - name: artifact-example
    steps:
    - - name: generate-artifact
        template: whalesay
    - - name: consume-artifact
        template: print-message
        arguments:
          artifacts:
          - name: message
            from: "{{steps.generate-artifact.outputs.artifacts.hello-art}}"

  - name: whalesay
    container:
      image: docker/whalesay:latest
      command: [sh, -c]
      args: ["cowsay hello world | tee /tmp/hello_world.txt"]
    outputs:
      artifacts:
      - name: hello-art
        path: /tmp/hello_world.txt

  - name: print-message
    inputs:
      artifacts:
      - name: message
        path: /tmp/message
    container:
      image: alpine:latest
      command: [sh, -c]
      args: ["cat /tmp/message"]
```

2）使用脚本任务产生的变量

使用脚本任务比容器任务更加灵活，通过脚本化的配置，可以极大地提升 Workflow 的使用效率。当任务执行完成后，可以在下一个任务中引用上一个任务输出的变量。通过这样的方式，我们能够用脚本编写 AI 模型中的预处理任务，直接配置在 Argo Workflow 的配置文件中。下面的配置示例展示了如何使用脚本任务及脚本任务产生的变量：

```yaml
apiVersion: argoproj.io/v1alpha1
kind: Workflow
metadata:
  generateName: scripts-bash-
spec:
  entrypoint: bash-script-example
  templates:
  - name: bash-script-example
    steps:
    - - name: generate
        template: gen-random-int-bash
    - - name: print
        template: print-message
        arguments:
          parameters:
          - name: message
            value: "{{steps.generate.outputs.result}}"

  - name: gen-random-int-bash
    script:
      image: debian:9.4
      command: [bash]
      source: |
        cat /dev/urandom | od -N2 -An -i | awk -v f=1 -v r=100 '{printf "%i\n", f + r * $1 / 65536}'

  - name: gen-random-int-python
    script:
      image: python:alpine3.6
      command: [python]
      source: |
        import random
        i = random.randint(1, 100)
```

```
        print(i)

- name: gen-random-int-javascript
  script:
    image: node:9.1-alpine
    command: [node]
    source: |
      var rand = Math.floor(Math.random() * 100);
      console.log(rand);

- name: print-message
  inputs:
    parameters:
    - name: message
  container:
    image: alpine:latest
    command: [sh, -c]
    args: ["echo result was: {{inputs.parameters.message}}"]
```

7.2.7　Traefik

1. 安装 Traefik

在 AI 平台中，我们会启动异常检测、时序预测、根因分析等多种 AI 推理服务，每个服务都有对外暴露的 API 接口，这时就需要使用 Kubernetes 的 Ingress 统一管理对外暴露的 API。

Traefik、API Six、Kong 都是非常优秀的 Ingress 组件，本节主要介绍 Traefik。

在 Kubernetes 中通过 Helm 能够快速地安装 Traefik：

```
helm repo add traefik https://helm.traefik.io/traefik
helm repo update
helm install traefik traefik/traefik
```

为了便于管理，编写 dashboard.yml，打开 Traefik 的 Dashboard：

```
apiVersion: traefik.containo.us/v1alpha1
kind: IngressRoute
```

```
metadata:
  name: dashboard
spec:
  entryPoints:
    - web
  routes:
  - match: Host(`traefik.localhost`) && (PathPrefix(`/dashboard`) || PathPrefix(`/api`))
      kind: Rule
      services:
        - name: api@internal
          kind: TraefikService
```

Traefik Dashboard 如图 7-10 所示。

图 7-10

通过 Traefik Dashboard，我们可以非常直观地看到系统当前的 Routes、Services 和 Middlewares 的情况。

2. Traefik 的关键概念

在 Kubernetes 中使用 Traefik，有 3 个需要我们了解的关键概念，分别是 Routes、Services、Middlewares。

Route 在 Traefik 中负责对流入的 Request 进行路由，引导 Request 到能够处理该 Request 的

Service，在引导 Request 的同时，若在 Route 中定义了 Middleware，则 Route 会在引导 Request 之前调用 Middleware。

常用的 Route 配置如表 7-7 所示。

表 7-7

规则	描述
Headers(`key`, `value`)	检查 Header 中的 Key 是否包含 Value
HeadersRegexp(`key`, `regexp`)	检查 Key 是否在 Header 中存在
Host(`example.com`, ...)	检查请求的域名是否为指定的域名
HostHeader(`example.com`, ...)	检查请求的域名中是否包含指定的 Header
HostRegexp(`example.com`, `{subdomain:[a-z]+}.example.com`, ...)	检查请求的域名是否匹配指定的正则表达式
Method(`GET`, ...)	检查请求的方式是否为指定的请求模式，如 GET、POST
Path(`/path`, `/articles/{cat:[a-z]+}/{id:[0-9]+}`, ...)	检查请求的路径是否为指定的路径，可使用正则表达式
PathPrefix(`/products/`, `/articles/{cat:[a-z]+}/{id:[0-9]+}`)	检查请求的路径是否以指定的路径开头
Query(`foo=bar`, `bar=baz`)	检查请求中是否带有指定的参数

在创建 Traefik Dashboard 时，我们就使用了 Rule，配置请求的域名以 traefik.localhost 开头，并且将请求路径以/dashboard 和/api 开头的请求转发至 api@internal 服务。

- match: Host(`traefik.localhost`) && (PathPrefix(`/dashboard`) || PathPrefix(`/api`))

Service 的概念相对简单，就是提供服务的后端程序。值得注意的是，在 Kubernetes 中，Traefik 的 Service 名称建议都填写为 Kubernetes 的 Service 名称。

Middleware 是 Traefik 一个非常重要的功能，它为我们提供了干预 Request 的机会。常用的 Middleware 如表 7-8 所示。

表 7-8

名称	描述
AddPrefix	为请求添加 Prefix
BasicAuth	为请求添加 HTTP Basic Auth
Buffering	为 request 和 response 添加缓存
Chain	串联多个 Middleware
CircuiteBeaker	为后端服务添加熔断器

续表

名称	描述
Compress	压缩 response 的结果
Headers	添加或更新 request 的 Header
IPWhiteList	添加发起 request 的 IP 地址白名单
RateLimit	为 request 添加限速器
ReplacePath	修改 request 的路径
StripPrefix	修改 request 的路径

7.3 小结

让系统具备 AIOps 能力的方案多种多样，但是还没有一个标准化、形成共识的落地方案，在这么多种落地方案中，通过 AI 平台赋能的落地方案是一种对运维系统侵入性较小、个性化较强、能够对 AI 能力进行统一管控的落地方案。AI 平台的初次建设成本相比起其他方案来说较大，但是建成之后能够让 AI 能力的生产速度得到大幅度的提升，适合有长期 AI 能力建设规划的团队。